Oxford Graduate Texts in N

Series Editors
R. Cohen   S.K. Donaldson   S. Hildebrandt
T.J. Lyons   M.J. Taylor

OXFORD GRADUATE TEXTS IN MATHEMATICS

Books in the series

1. Keith Hannabuss: *An introduction to quantum theory*
2. Reinhold Meise and Dietman Vogt: *Introduction to functional analysis*
3. James G. Oxley: *Matroid theory*
4. N.J. Hitchin, G.B. Segal, and R.S. Ward: *Integrable systems: twistors, loop groups, and Rieman surfaces*
5. Wulf Rossmann: *Lie groups: An introduction through linear groups*
6. Qing Liu: *Algebraic geometry and arithmetic curves*
7. Martin R. Bridson and Simon M. Salamon (eds): *Invitations to geometry and topology*
8. Shumel Kantotovitz: *Introduction to modern analysis*
9. Terry Lawson: *Topology: A geometric approach*
10. Meinolf Geck: *An introduction to algebraic geometry and algebraic groups*
11. Alastair Fletcher and Vladimir Markovic: *Quasiconformal maps and Teichmüller theory*
12. Dominic Joyce: *Riemannian holonomy groups and calibrated geometry*
13. Fernando Rodriguez Villegas: *Experimental Number Theory*

Experimental Number Theory

Fernando Rodriguez Villegas

OXFORD
UNIVERSITY PRESS

OXFORD
UNIVERSITY PRESS

Great Clarendon Street, Oxford OX2 6DP

Oxford University Press is a department of the University of Oxford.
It furthers the University's objective of excellence in research, scholarship,
and education by publishing worldwide in

Oxford New York

Auckland Cape Town Dar es Salaam Hong Kong Karachi
Kuala Lumpur Madrid Melbourne Mexico City Nairobi
New Delhi Shanghai Taipei Toronto

With offices in

Argentina Austria Brazil Chile Czech Republic France Greece
Guatemala Hungary Italy Japan Poland Portugal Singapore
South Korea Switzerland Thailand Turkey Ukraine Vietnam

Oxford is a registered trade mark of Oxford University Press
in the UK and in certain other countries

Published in the United States
by Oxford University Press Inc., New York

© Oxford University Press, 2007

The moral rights of the author have been asserted
Database right Oxford University Press (maker)

First published 2007

All rights reserved. No part of this publication may be reproduced,
stored in a retrieval system, or transmitted, in any form or by any means,
without the prior permission in writing of Oxford University Press,
or as expressly permitted by law, or under terms agreed with the appropriate
reprographics rights organization. Enquiries concerning reproduction
outside the scope of the above should be sent to the Rights Department,
Oxford University Press, at the address above

You must not circulate this book in any other binding or cover
and you must impose the same condition on any acquirer

British Library Cataloguing in Publication Data

Data available

Library of Congress Cataloging in Publication Data

Data available

Typeset by Newgen Imaging Systems (P) Ltd., Chennai, India
Printed in Great Britain
on acid-free paper by
Biddles Ltd., King's Lynn, Norfolk

ISBN 978–0–19–852822–7
ISBN 978–0–19–922730–3 (pbk)

1 3 5 7 9 10 8 6 4 2

To Lula, Malena and Adriana

Acknowledgements

Numerous people have contributed to this book, in one way or another. I thank them all.

First and foremost, I would like to thank Karim Belabas, the main maintainer of PARI-GP, who read the book line by line and suggested many significant improvements both in the text and in the coding of the scripts. His help was invaluable (remaining bugs are of course mine). I would also like to thank: Don Zagier, from whom I learned, over the years, a lot of mathematics both theoretical and experimental; John Tate, for our many conversations, mathematical and otherwise, and for his many suggestions and comments on the manuscript; David Boyd, experimental number theorist *par excellence*, whose work I used in many examples in the book; Paul Gunnells, for a careful reading of an earlier version; and Margaret Combs, for the figures in the book and her general TeXpert advice.

Special thanks go to my family for their love.

Preface

In May 30, 1799 Gauss wrote in his mathematical diary

Terminum medium arithmetico–geometricum inter 1 et $\sqrt{2}$ esse = $\frac{\pi}{\varpi}$ usque ad figuram undecimam comprobavimus, qua re demonstrata prorsus novus campus in analysis certo aperietur.

which translates to (see [27] p. 281)

We have established that the arithmetic–geometric mean between 1 and $\sqrt{2}$ is $\frac{\pi}{\varpi}$ to the eleventh decimal place; the demonstration of this fact will surely open an entirely new field of analysis.

The arithmetic–geometric mean $M(a, b)$ of two positive real numbers a, b is the common limit of the sequence a_n, b_n defined by the recursion

$$a_{n+1} := \tfrac{1}{2}(a_n + b_n), \quad b_{n+1} := \sqrt{a_n b_n}, \quad a_0 := a, \quad b_0 := b. \quad (0.1)$$

The symbol ϖ was Gauss's special notation for the integral

$$\varpi := 2 \int_0^1 \frac{1}{\sqrt{1 - x^4}} \, dx, \quad (0.2)$$

which had appeared in a 1691 paper of J. Bernoulli; the numerical value

$$\frac{\pi}{\varpi} = 2 \int_0^1 \frac{x^2}{\sqrt{1 - x^4}} \, dx = 1.19814023473559223\ldots \quad (0.3)$$

was well known to Gauss after its calculation to 17 decimals by J. Stirling in 1730.

There is no doubt that Gauss's remark was prophetic (see [27] for a beautiful modern rendition of this *entirely new field* that he was to discover). In fact, numerical experimentation is crucial to number theory, perhaps more so than to other areas of mathematics. Witness the Birch–Swinnerton-Dyer conjecture, one of the outstanding million-dollar millennium problems of the Clay Mathematics Institute, which was inferred from numerical calculations. Indeed, as Cassels has said, to a large degree number theory is an experimental science.

Here is what Hermite wrote to Stieltjes in his letter of March 12, 1884 (in my translation from the french original, see [4] letter 47)

The difficulty is in recognizing in which way such different expressions are related to one another; having no hope to surmount it, I resign myself, Sir, to question the different kinds of elliptic formulas, asking to each one its arithmetic secret, and to collect the useful or useless answers with patience and perserverance; *plus laboris quam artis*.

The goal of this book is to introduce the reader to the use of the computer as a research tool in number theory. I will discuss a variety of different examples to illustrate some of what the current state of software and hardware allows us to compute on an everyday basis. The discussion will not be concerned with the most clever possible approach to a given calculation or a careful analysis of the running time of a given algorithm. Instead, it will concentrate on what works on a small to medium scale; say, what can be done, from devising the algorithm, coding it, running it to analyzing its output, in under an hour. In other words, I will consider human time as part of the running time of an algorithm: let us start the clock when we first think of the question and stop it when the computer finally produces the output.

I have not attempted the impossible task of being comprehensive, but have tried to cover a broad spectrum of basic computational issues. The only way I can personally learn how to use a new piece of software is by analyzing many worked out examples. That is the approach taken here. My hope is, also, that bits and pieces of the examples in the book could be of use for your own mathematical questions.

All of the programs in this book are written in GP, the scripting language for the computational package PARI which I have been using for 15 years. The GP scripts and examples have been tested using the stable release version 2.3.0. They are available for download at

http://www.math.utexas.edu/users/villegas/gpbook

The input and output in the examples has been lightly reformatted for the book so they might look different than in an actual GP session on a computer.

For the basics of GP and description of its built-in functions we refer the reader to its manual and tutorial, which can be found at the main site

http://pari.math.u-bordeaux.fr/

The book, however, is definitely *not* a PARI manual.

Except for those in Chapter 5 on combinatorics the routines in the book are all pretty short and straightforward. Breaching standard programming etiquette I have not included any comments within their code. Then again, I personally find that attempting to decipher a simple text in an unknown (human) language is the most useful and rewarding way to learn it.

The book is also not an introduction to number theory though there is quite a bit of theory blended in. I will assume the reader has a basic knowledge of the arithmetic of number fields, Galois theory, p-adic numbers, etc. However, going through some the examples in the book, in combination with a more standard text as a reference, could be of great help in learning this material.

I know bugs are lurking in what follows and in this regard, I cannot resist one last quote. In a letter to Euler (December 24, 1742) where Goldbach discusses what have now become know as *multi-zeta values*, he mentions that the series arose from a slip of the pen on his part and brings up the saying *Si non errasset, fecerat ille minus*. This is a line from a poem by the roman poet M. V. Martialis which translates into something like *if he had not erred, he would have done less*.

Of course, no Latin verse can assuage the frustration of a numerical experiment gone wrong. The most remarkable computer bug story I have heard happened to M. Rubinstein. In a certain calculation to check the results on a paper with B. Poonen [76] things worked to a certain point and then mysteriously broke down. It took Rubinstein three days to figure out that the bug was in the high-precision approximation to π he was using. He got it from cutting and pasting it from another window in his computer screen. In the process he had left out one digit somewhere in the middle!

One last thing: I anticipate that reading this book without a computer at hand would be rather useless. Try out the routines as you read and, by all means, experiment!

Contents

1	**Basic examples**	**1**
1.1	How things vary with p	1
	1.1.1 Quadratic Reciprocity Law	1
	1.1.2 Sign functions	5
	1.1.3 Checking modularity of sign functions	6
	1.1.4 Examples	7
1.2	Recognizing numbers	10
	1.2.1 Rational numbers in **R**	10
	1.2.2 Rational numbers modulo m	14
	1.2.3 Algebraic numbers	15
1.3	Bernoulli polynomials	19
	1.3.1 Definition and properties	19
	1.3.2 Calculation	20
	1.3.3 Related questions	21
	1.3.4 Proofs	23
1.4	Sums of squares	27
	1.4.1 $k=1$	28
	1.4.2 $k=2$	28
	1.4.3 $k \geq 3$	29
	1.4.4 The numbers $r_k(n)$	31
	1.4.5 Theta functions	36
1.5	Exercises	36
2	**Reciprocity**	**40**
2.1	More variation with p	40
	2.1.1 Local zeta functions	40
	2.1.2 Formulation of reciprocity	41
	2.1.3 Global zeta functions	43

2.2　The cubic case　　　　　　　　　　　　　　　　　　　　44
　　　　　　2.2.1　Two examples　　　　　　　　　　　　　　　　45
　　　　　　2.2.2　A modular case　　　　　　　　　　　　　　　46
　　　　　　2.2.3　A non-modular case　　　　　　　　　　　　　48
　　　2.3　The Artin map　　　　　　　　　　　　　　　　　　　　50
　　　　　　2.3.1　A Galois example　　　　　　　　　　　　　　50
　　　　　　2.3.2　A non-Galois example　　　　　　　　　　　　53
　　　2.4　Quantitative version　　　　　　　　　　　　　　　　　55
　　　　　　2.4.1　Application of a theorem of Tchebotarev　　　　55
　　　　　　2.4.2　Computing densities numerically　　　　　　　56
　　　2.5　Galois groups　　　　　　　　　　　　　　　　　　　　59
　　　　　　2.5.1　Tchebotarev's theorem　　　　　　　　　　　　63
　　　　　　2.5.2　Trink's example　　　　　　　　　　　　　　　64
　　　　　　2.5.3　An example related to Trink's　　　　　　　　　65
　　　2.6　Exercises　　　　　　　　　　　　　　　　　　　　　　69

3　Positive definite binary quadratic forms　　　　　　　　　　71

　　　3.1　Basic facts　　　　　　　　　　　　　　　　　　　　　71
　　　　　　3.1.1　Reduction　　　　　　　　　　　　　　　　　　72
　　　　　　3.1.2　Cornachia's algorithm　　　　　　　　　　　　73
　　　　　　3.1.3　Class number　　　　　　　　　　　　　　　　75
　　　　　　3.1.4　Composition　　　　　　　　　　　　　　　　　76
　　　3.2　Examples of reciprocity for imaginary quadratic fields　　77
　　　　　　3.2.1　Dihedral group of order 6　　　　　　　　　　　77
　　　　　　3.2.2　Theta functions again　　　　　　　　　　　　79
　　　　　　3.2.3　Dihedral group of order 10　　　　　　　　　　82
　　　　　　3.2.4　An example of F. Voloch　　　　　　　　　　　84
　　　　　　3.2.5　Final comments　　　　　　　　　　　　　　　90
　　　3.3　Exercises　　　　　　　　　　　　　　　　　　　　　　91

4　Sequences　　　　　　　　　　　　　　　　　　　　　　　　92

　　　4.1　Trinomial numbers　　　　　　　　　　　　　　　　　　92
　　　　　　4.1.1　Formula　　　　　　　　　　　　　　　　　　　92
　　　　　　4.1.2　Differential equation and linear recurrence　　　93
　　　　　　4.1.3　Algebraic equation　　　　　　　　　　　　　　98
　　　　　　4.1.4　Hensel's lemma and Newton's method　　　　　99
　　　　　　4.1.5　Continued fractions　　　　　　　　　　　　　103
　　　　　　4.1.6　Asymptotics　　　　　　　　　　　　　　　　　107

		4.1.7 More coefficients in the asymptotic expansion	109
		4.1.8 Can we sum the asymptotic series?	111
	4.2	Recognizing sequences	113
		4.2.1 Values of a polynomial	113
		4.2.2 Values of a rational function	114
		4.2.3 Constant term recursion	115
		4.2.4 A simple example	117
	4.3	Exercises	119

5	Combinatorics	123
	5.1 Description of the basic algorithm	123
	5.2 Partitions	127
	5.2.1 The number of partitions	128
	5.2.2 Dual partition	130
	5.3 Irreducible representations of S_n	131
	5.3.1 Hook formula	132
	5.3.2 The Murnaghan–Nakayama rule	133
	5.3.3 Counting solutions to equations in S_n	138
	5.3.4 Counting homomorphism and subgroups	139
	5.4 Cyclotomic polynomials	143
	5.4.1 Values of ϕ below a given bound	143
	5.4.2 Computing cyclotomic polynomials	145
	5.5 Exercises	148

6	*p*-adic numbers	151
	6.1 Basic functions	151
	6.1.1 Mahler's expansion	151
	6.1.2 Hensel's lemma and Newton's method (again)	152
	6.2 The *p*-adic gamma function	155
	6.2.1 The multiplication formula	158
	6.3 The logarithmic derivative of Γ_p	159
	6.3.1 Application to harmonic sums	161
	6.3.2 A formula of J. Diamond	163
	6.3.3 Power series expansion of $\psi_p(x)$	164
	6.3.4 Application to congruences	167

	6.4	Analytic continuation	168
		6.4.1 An example of Dwork	169
		6.4.2 A generalization	170
		6.4.3 Dwork's exponential	170
	6.5	Gauss sums and the Gross–Koblitz formula	172
		6.5.1 The case of $\mathbf{F}_p$	172
		6.5.2 An example	174
	6.6	Exercises	175

7 Polynomials 177

	7.1	Mahler's measure	177
		7.1.1 Simple search	178
		7.1.2 Refining the search	179
		7.1.3 Counting roots on the unit circle	182
	7.2	Applications of the Graeffe map	182
		7.2.1 Detecting cyclotomic polynomials	182
		7.2.2 Detecting cyclotomic factors	184
		7.2.3 Wedge product polynomial	184
		7.2.4 Interlacing roots of unity	185
	7.3	Exercises	188

8 Remarks on selected exercises 189

References 205

Index 211

1 Basic examples

In this chapter we will explore some basic examples, introducing mathematical and computational concepts that we will develop further later.

1.1 How things vary with *p*

Consider a system of congruence equations

$$\begin{cases} F_1(x_1,\ldots,x_n) \equiv 0 \bmod m \\ \quad\vdots \\ F_N(x_1,\ldots,x_n) \equiv 0 \bmod m \end{cases} \quad (1.1)$$

where the F_i's are polynomials with integer coefficients. This system has finitely many solutions for a fixed m and it is natural to consider their number $N(m)$. By the chinese remainder theorem $N(mm') = N(m)N(m')$ if m and m' are relatively prime and hence, without loss of generality, we may consider only $N(p^r)$ where p is prime. Moreover, as we will argue later §6.1.2, typically $N(p)$ alone determines $N(p^r)$ for all other r; therefore, in this chapter we will only consider equations modulo a prime number p. One major advantage of this assumption is that the integers modulo a prime p form a field (see Ex. 1.5).

How does $N(p)$ vary with p? Already in the first non-trivial case of a quadratic equation in one variable, this is a subtle question. Indeed, the answer is given by the quadratic reciprocity law.

1.1.1 Quadratic Reciprocity Law

Recall that a *Dirichlet character* χ modulo m is a group homomorphism

$$\chi : (\mathbf{Z}/m\mathbf{Z})^* \longrightarrow \mathbf{C}^*. \quad (1.2)$$

Following a common abuse of notation we also denote by χ the function on $\mathbf{Z}$ with values in $\mathbf{C}$ defined by $a \mapsto \chi(a \bmod m)$ if $\gcd(a,m) = 1$ and

$a \mapsto 0$ otherwise. As a function on $\mathbf{Z}$ it satisfies $\chi(mn) = \chi(m)\chi(n)$ for all $m, n \in \mathbf{Z}$ and is therefore uniquely determined by its values $\chi(p)$ for primes p and $\chi(-1)$.

A character χ modulo m can be regarded as a character χ'' modulo any multiple m'' of m by composing with the reduction map

$$(\mathbf{Z}/m''\mathbf{Z})^* \longrightarrow (\mathbf{Z}/m\mathbf{Z})^*. \tag{1.3}$$

Viewed as functions on $\mathbf{Z}$, χ and χ'' agree on all but finitely many primes. On the other hand, the character χ could itself arise from a character χ' modulo a divisor m' of m in this way. The minimal such m' is called the *conductor* of χ and χ' is its *associated primitive character*. We denote the conductor of χ by $\mathrm{cond}(\chi)$. We say that χ is *primitive* if its conductor is exactly m (and hence $\chi' = \chi$). The *trivial character* is the character that takes the value 1 on any non-zero integer; it is the unique character of conductor 1.

For example, there are two primitive characters of conductor $m = 8$; they are given by the Kronecker symbols $(\frac{\pm 8}{\cdot})$. We can compute the values of these characters with GP

```
? vector(8,n,kronecker(8,n-1))

[0, 1, 0, -1, 0, -1, 0, 1]

? vector(8,n,kronecker(-8,n-1))

[0, 1, 0, 1, 0, -1, 0, -1]
```

Their product has values

```
[0, 1, 0, -1, 0, 1, 0, -1]
```

which is not a primitive character modulo 8; its associated primitive character has conductor 4 and is given by the Kronecker symbol $(\frac{-4}{\cdot})$.

It will be convenient to introduce the following notation. A *(finite) étale algebra* over $\mathbf{Q}$ is a $\mathbf{Q}$-algebra of the form

$$A = \mathbf{Q}[x]/(f),$$

where $f \in \mathbf{Q}[x]$ is a polynomial with non-zero discriminant. Note that the assumption on the discriminant is equivalent to f factoring into distinct irreducible factors

$$f = f_1 \cdots f_r$$

over Q. We then have

$$A = \prod_{i=1}^{r} K_i,$$

where each $K_i = \mathbf{Q}[x]/(f_i)$ is a *number field* (a finite field extension of Q).

We define the *degree* $[A:\mathbf{Q}]$ of A as its dimension as a vector space over Q, which is equal to the degree of f. We call an A of degree 2 quadratic for short. Of course, the algebra A does not determine the polynomial f uniquely; we should think of (the isomorphism class) of A as the intrinsic object and f as one of many possible equations for it.

For example, for quadratic algebras if $f = ax^2 + bx + c$ and $D := b^2 - 4ac = \mathrm{disc}(f)$ then

$$\mathbf{Q}(\sqrt{D}) \longrightarrow \mathbf{Q}[x]/(f)$$
$$\sqrt{D} \mapsto 2ax + b,$$

where $\mathbf{Q}(\sqrt{D}) := \mathbf{Q}[x]/(x^2 - D)$ is an isomorphism by the quadratic formula. It is easy to check that $\mathbf{Q}(\sqrt{D_1})$ and $\mathbf{Q}(\sqrt{D_2})$ are isomorphic if and only if $D_1 = z^2 D_2$ for some $z \in \mathbf{Q}^\times$. Hence, the isomorphism classes of étale quadratic algebras over Q are parameterized by $\mathbf{Q}^\times/(\mathbf{Q}^\times)^2$. (For $D = 0$ we have the algebra of *dual numbers* $\mathbf{Q}[x]/(x^2)$ which is not étale.)

It turns out, however, that equations of the form $x^2 - D$ for quadratic algebras are not that convenient. We define instead a *global minimal model* of A to be a polynomial $f = x^2 + bx + c$ with $b, c \in \mathbf{Z}$ such that $A \simeq \mathbf{Q}[x]/(f)$ and $|\mathrm{disc}(f)|$ is minimal. (WARNING: this is a bit of a hack which works well for quadratic algebras but *not* for algebras of higher degree, see §2.1.2.) We call $\mathrm{disc}(f)$ for a minimal model the *discriminant* of A and denote it $\mathrm{disc}(A)$. The isomorphism class of A is uniquely determined by $\mathrm{disc}(A)$.

For example, the algebra $A = \mathbf{Q}(\sqrt{-3})$ has minimal model $f = x^2 + x + 1$ and $\mathrm{disc}(A) = -3$, whereas $A = \mathbf{Q}(\sqrt{3})$ has minimal model $x^2 - 3$ and $\mathrm{disc}(A) = 12$.

The discriminant $D := b^2 - 4ac = \mathrm{disc}(f)$ of any polynomial $f = ax^2 + bx + c \in \mathbf{Z}[x]$ with $A \simeq \mathbf{Q}[x]/(f)$ is of the form $z^2 \mathrm{disc}(A)$ for some non-zero integer z. Note that $D \equiv 0, 1 \bmod 4$; we call any non-zero integer D with this property a *discriminant*. (It is easy to check that any discriminant is indeed the discriminant of such an f.)

We say that D is a *fundamental discriminant* if it is the discriminant of a quadratic étale algebra. (In GP, we may test this with `isfundamental(D)`, which returns 1 if true and 0 if not.) Let us note that $D = 1$ is a fundamental discriminant; it corresponds to the only quadratic étale algebra which is not a field, namely, $\mathbf{Q} \times \mathbf{Q}$.

Here is a short list of fundamental discriminants computed with GP.

```
? for(D=-20,20,if(isfundamental(D),print1(D," ")))

-20 -19 -15 -11 -8 -7 -4 -3 1 5 8 12 13 17
```

One way to think of the quadratic reciprocity is that it establishes a one to one correspondence $A \longleftrightarrow \chi_A$, between isomorphism classes of quadratic étale algebras and primitive Dirichlet characters of order dividing 2. Under this correspondence we have $\text{cond}(\chi_A) = |\text{disc}(A)|$ and

$$N_A(p) = \chi_A(p) + 1 \tag{1.4}$$

for all primes p, where $N_A(p)$ is the number of solutions to the congruence $f(x) \equiv 0 \bmod p$ with f a global minimal model of A.

If $f \in \mathbb{Z}[x]$ is an arbitrary equation defining A then (1.4) is valid for all but finitely many primes. In particular, we may take the equation $f = x^2 - \text{disc}(A)$ and give (1.4) in the more familiar form in terms of quadratic symbols

$$\chi_A(p) = \left(\frac{\text{disc}(A)}{p}\right). \tag{1.5}$$

This is actually valid for all primes p (using the *Kronecker symbol* on the right hand side).

For example, if $A = \mathbb{Q}(\sqrt{\pm 2})$ then $\text{disc}(A) = \pm 8$ and χ_A are the two primitive characters mentioned above.

For completeness let us check a numerical example by brute force. Say $f = x^2 + x - 1$ a global minimal model for the algebra $A = \mathbb{Q}(\sqrt{5})$. First we need a function that will compute $N(p)$

```
count(f,p)=sum(a=0,p-1,subst(f,x,Mod(a,p))==0)
```

Note the somewhat unusual mixture of boolean and standard arithmetic: the term in the sum (subst(f,x,Mod(a,p))==0) is 1 if $f(a) \equiv 0 \bmod p$ and is 0 otherwise.

```
? forprime(p=2,50,print(p, "\t", count(x^2+x-1,p), "\t",
kronecker(5,p)+1))
```

2	0	0
3	0	0
5	1	1
7	0	0
11	2	2
13	0	0
17	0	0
19	2	2

23	0	0
29	2	2
31	2	2
37	0	0
41	2	2
43	0	0
47	0	0

In a weak form the quadratic reciprocity law says that the number of solutions to the congruence $f(x) \equiv 0 \bmod p$ for $f \in \mathbf{Z}[x]$ of degree 2 is completely determined by the class of p modulo a number m that depends only on f.

What about the value $\chi_A(-1)$? It corresponds to the prime ∞; i.e., to the real numbers. Consider $A \otimes_\mathbf{Q} \mathbf{R}$; it is a quadratic étale algebra over $\mathbf{R}$ and hence it must be isomorphic to either $\mathbf{C}$ or $\mathbf{R} \times \mathbf{R}$ depending on whether $\operatorname{disc}(A) < 0$ or $\operatorname{disc}(A) > 0$, respectively. We may write this as $N(\infty) = 1 + \chi_A(-1)$ or simply

$$\chi_A(-1) = \operatorname{sgn}(\operatorname{disc}(A)).$$

1.1.2 Sign functions

We call any function Φ sending primes to $\{-1, 0, 1\}$ a *sign function*. We say that a sign function is *modular* if $\Phi(p) = \chi(p)$ for some primitive Dirichlet character χ for all but finitely many p (there is no loss of generality in assuming that χ is primitive as we are willing to ignore a finite set of primes). If m is the conductor of χ we say that Φ is modular of conductor m. For example, we may state the quadratic reciprocity law as saying that for any quadratic étale algebra A the map $p \mapsto N_A(p) - 1$ is modular with character χ_A.

We will discuss in more detail what happens with $N(p)$ for one-variable polynomials of higher degree in the next chapter. For the moment, however, consider the following sign function. Given a monic polynomial $f \in \mathbf{Z}[x]$ with non-zero discriminant D define

$$\Phi_f(p) = \begin{cases} (-1)^{\sum_h (\deg(h)+1)} & \text{if } p \nmid D \\ 0 & \text{if } p \mid D \end{cases} \quad (1.6)$$

where $f \equiv \prod_h h \bmod p$ is the factorization of f into irreducible factors modulo p. (The value of $\Phi_f(p)$ for $p \nmid D$ is the sign of any permutation in the Frobenius conjugacy class Frob_p see §2.3.)

The quadratic reciprocity law implies that Φ_f is modular for f quadratic. This is in fact true in general.

Theorem 1.1 *Let $f \in \mathbb{Z}[x]$ be monic with non-zero discriminant D and let Φ_f be the sign function defined above (1.6). Then Φ_f is modular with Dirichlet character that associated to $\mathbb{Q}(\sqrt{D})$.*

1.1.3 Checking modularity of sign functions

Suppose we are given a sign function Φ, how do we test numerically whether it is modular? We may use the following GP script, which we analyze in some detail.

```
testmod(v,bd)=
{
  local(n,mn,S);

  mn=length(v); S=[0,mn];
    for(D=-bd,bd,
      if(!isfundamental(D),next);
        n=dist(v,D);
          if(n < mn,
            mn=n;
            S=[D,mn]));
  S
}
```

The input consists of a vector v and a bound bd, a positive integer. The vector v is a sampling of the function Φ we want to test for modularity; its entries are themselves vectors of length 2 of the form (p, ϵ_p) where p is an integer and $\Phi(p) = \epsilon_p \in \{-1, 0, 1\}$. (According to our definition, and typically, p will be prime but this is not strictly necessary.)

The output is a vector (D, n) where D is the fundamental discriminant in the given range that minimizes the distance

$$n(D) = \sum_p |\epsilon_p - \chi(p)|, \qquad (1.7)$$

where χ is the primitive quadratic character associated to D.

We run over all integers from -bd to bd (the loop for(D=-bd, bd, ...)) and use only those D's which are fundamental (testing with isfundamental). We compute the distance n of our input to the Kronecker symbol at D and compare to the current minimum mn; if it is smaller we make the new minimum and record in S which discriminant it corresponds to. At the end of the loop we have a two component vector S of the form $[D, n]$ with D a fundamental discriminant that achieves the minimum distance n within the given range of discriminants. Note that the output is one of possibly many discriminants that achieve this minimum distance.

The routine calls the function

```
dist(v,D)=
{
  sum(j=1,length(v),
    abs(v[j][2]-kronecker(D,v[j][1])))
}
```

which given a vector `v` as in the input for `testmod` and an integer D computes the distance n defined in (1.7). Note how entries in a vector are retrieved: `v[j][1]` is the first entry of the two-component vector which is the j-th entry of the vector `v`, etc.

As with any exploratory test, one should be careful to interpret the output. If, for example, Φ is modular but we give `bd` less than its conductor the output will be meaningless; we should also make sure we give enough data points of Φ. As a rule of thumb, we may have some confidence in the answer if it does not change when we successively increase `bd` (more discriminants to test against) and the length of `v` (more data). Even then one should not jump to conclusions too soon, our data might be biased, consisting, for example, only of primes which are quadratic residues modulo some fixed number, etc.

Note that n need not be zero as $\Phi(p)$ and $\chi(p)$ are allowed to differ at finitely many primes but certainly a small n relative to `length(v)` is a sign that the output probably is the right answer. There is also another reason why it is a good idea to have the flexibility of getting an answer even if we do not have a perfect matching of $\Phi(p)$ and $\chi(p)$: the data could be corrupted (due to a cut and pasting or typing error say; see the Preface for a particularly nasty example). In that case, getting no answer would not be of any help.

1.1.4 Examples

Here is a GP script to compute the sign function Φ_f we associated (1.6) to polynomials.

```
sgn(f,p)=
{
  if(poldisc(f)%p,return(0));
    f=factormod(f,p);
    (-1)^sum(j=1,length(f~),poldegree(f[j,1])+1)
}
```

This is a literal translation of the mathematical definition (1.6). The conditional `if(poldisc(f)%p,...,...)` tests whether p divides the discriminant D of the polynomial f by computing the remainder of the division of D by p. If the remainder is non-zero (first option) we first compute the factorization of f mod p with f=factormod(f,p) and then the expression in (1.6) in terms of the factors, otherwise (second option) we return zero.

Note how we recycled variables by assigning the result of factormod(f,p) to f itself. This modification is local to the function sgn and does not affect the actual input polynomial. In other words, if we had previously defined say f=2*x+1 and then do sgn(f,5) the value of f would not be changed.

We can also improve the code by using the feature of factormod that returns the degrees of the factors without the factors themselves (which we do not actually use). See polfacttype in §2.2 for an example.

As an illustration let us compute the sign function for the fifth cyclotomic polynomial $f = x^4 + x^3 + x^2 + x + 1$ (given in GP by polcyclo(5)) and all primes $p \leq 50$. Suppose that all the relevant functions are in a file mod-signs.gp in the current directory where GP is running. We start by reading this file into GP by typing

```
? \r mod-signs.gp
```

Then we type, for example,

```
? v=[]; forprime(p=2,50,
v=concat(v,[[p,sgn(polcyclo(5),p)]]));v
```

at the prompt and get

```
[[2, -1], [3, -1], [5, 0], [7, -1], [11, 1], [13, -1],
[17, -1], [19, 1], [23, -1], [29, 1], [31, 1], [37, -1],
[41, 1], [43, -1], [47, -1]]
```

We test with

```
? testmod(v,50)
```

getting the answer

```
[5, 0]
```

This of course agrees with Theorem 1.1 since the discriminant of f is 5^3 and $\mathrm{disc}(\mathbb{Q}(\sqrt{5^3})) = 5$.

For our next example we consider the following sign function.

$$\Phi(p) \equiv \begin{cases} \sum_{n=0}^{p-1} a_n \, 432^{-n} \bmod p & \text{if } p \nmid 6 \\ 0 & \text{if } p \mid 6, \end{cases} \quad (1.8)$$

where
$$a_n = \frac{(6n)!}{n!(2n)!(3n)!} \tag{1.9}$$

(it is a non-trivial fact that the sum on the right hand is always ± 1 modulo p). In GP we can define this as follows

```
sgn1(p) =
{
  if(6%p == 0, return(0));
    centerlift(Mod(1,p)*sum(n=0,p-1,
      (6*n)!/n!/(2*n)!/(3*n)!/432^n))
}
```

The combination `centerlift(Mod(1,p)*x)` gives an integer y in the range $|y| < p/2$ (for $p > 2$) with $x \equiv y \mod p$ (as opposed to simply `lift` which gives such a y in the range $0 \le y < p$).

Here is the output of this function for $2 \le p \le 50$.

```
? v=[]; forprime(p=2,50, v=concat(v,[[p,sgn1(p)]]));v

[[2, 0], [3, 0], [5, 1], [7, -1], [11, -1], [13, 1],
 [17, 1], [19, -1], [23, -1], [29, 1], [31, -1], [37, 1],
 [41, 1], [43, -1], [47, -1]].
```

We test for modularity

```
? testmod(%3,40)
[-4, 1]
```

(We naturally could have guessed this directly by staring at the list of values for a few minutes.)

It can be proved that indeed $\Phi(p) = \left(\frac{-4}{p}\right)$ for all $p > 3$. In fact, numerically it appears that the stronger congruence

$$\sum_{n=0}^{p-1} \frac{(6n)!}{n!(2n)!(3n)!} 432^{-n} \equiv \left(\frac{-4}{p}\right) \mod p^2$$

holds; this has been recently proved by E. Mortenson [65].

Let us point out that if for some reason we really wanted to compute $\Phi(p)$ for large primes p using the definition, `sgn1` would be rather inefficient. We would be summing a not-so-large ratio of some very big numbers $((6n)!, (3n)!, \text{etc.})$ and then reducing the result modulo p. We are better off computing the numbers a_n recursively by noting that

$$\frac{a_{n+1}}{a_n} = 12 \frac{(6n+1)(6n+5)}{(n+1)^2}. \tag{1.10}$$

The new function would then be something like

```
sgn2(p) =
{
  local(M, S);

    if (6%p == 0, return (0));
      S = M = Mod(1,p);
      for (n=1, p-1,
        S += (M *= (6*n-1)*(6*n-5) / (n^2*36)));
    centerlift(S)
}
```

Since we only want the value of the sum modulo p we do the recursion modulo p (note that the denominator is never zero modulo p). This is quicker since it uses modular rather than integer arithmetic; we force GP to use arithmetic modulo p by simply initializing the first term to be 1 mod p, as opposed to just 1, with a=Mod(1,p).

1.2 Recognizing numbers

A common theme underlying this book is the numerical recognition of mathematical objects and patterns. We discuss here how to recognize rational and algebraic numbers from a given approximation.

1.2.1 Rational numbers in R

If we compute

```
? zeta(2)/Pi^2

0.16666666666666666666666666667
```

zeta is Riemann's zeta function (2.8) we do not need much to guess that $\zeta(2) = \pi^2/6$. This is, of course, a well known fact first proved by Euler. In fact, Euler proved that for any *even* integer n the ratio $\zeta(n)/\pi^n$ is a rational number.

Let us continue

```
? zeta(4)/Pi^4

0.01111111111111111111111111111
```

We may still quickly guess that $\zeta(4) = \pi^4/90$ but how about the next case?

```
? z=zeta(12)/Pi^12

0.0000010822021404031986042568053315
```

1.2 Recognizing numbers

Is $\zeta(12)/\pi^{12}$ really rational? How do we recognize what number it is from its approximation z?

If never faced with this question before the reader may be surprised to find out that we *do not* look for periodicity in the decimal expansion of the approximation. Instead we used the continued fraction algorithm. Consider the following GP script

```
recognize(x)=local(m);m=contfracpnqn(contfrac(x)); m[1,1]/m[2,1]
```

We assume the input is a real number x. We first compute the continued fraction of x with contfrac, which returns a vector with as many partial quotients [c0,c1,...,cn] as it can determine given the precision of x, and then output the last convergent p_n/q_n of this continued fraction. WARNING: the output of contfrac is normalized in the standard way with $c_n > 1$ unless $n = 0$ (to avoid duplications due to the formal identity $[c_0, c_1, \ldots, c_n-1, 1] = [c_0, c_1, \ldots, c_n]$); in particular, c_n may actually change if we increase the precision of x and recompute; the previous ones will not.

Let us try our number

```
? recognize(z)
```

691/638512875

This is indeed the correct value. What Euler really proved is the identity

$$\zeta(2n) = \frac{2^{2n-1}\pi^{2n}}{(2n)!}|B_{2n}|, \quad n = 1, 2, \ldots, \tag{1.11}$$

where $B_{2n} \in \mathbb{Q}$ is the *Bernoulli number* (see §1.3) and

```
? abs(bernfrac(12))*2^11/12!
```

691/638512875

In practice we should increase the precision of our calculation and run recognize again to see if the answer remains the same. We should emphasize that, of course, there is no way to *prove* with a computer that an unknown real number is rational no matter how good an approximation of it we have.

We should clarify what is the output of contfrac. If the input is rational then the continued fraction terminates and we get it exactly. If the input is a real number x only known to n significant digits (i.e., we know rational numbers a, b such that $a \leq x \leq b$ with $0 \leq b/a - 1 \leq 10^{-n}$) then contfrac returns those partial quotients than can be correctly computed with this knowledge alone. (For a random number we expect to get 0.97 partial quotients per decimal digit, see [52] vol. 2, 4.5.4 Ex. 47.)

1 : Basic examples

To illustrate this take z as above

```
? z=zeta(12)/Pi^12
```

0.000001082202140403198604256805315

and write it as a rational number

```
? zz=1082202140403198604256805315/10^33
```

216440428080639720851361063/2000000000000000000000000000000000
```
? zz*1.
```

0.000001082202140403198604256805315
```
? z-zz
```

1.953125000 E-35

Now let us see what happens when we compute their continued fractions.

```
? contfrac(z)
```

[0, 924041, 1, 3, 1, 2, 2, 1, 14]

```
? contfrac(zz)
```

[0, 924041, 1, 3, 1, 2, 2, 1, 13, 1, 384859869703995923, 3, 10, 3, 1, 1, 1, 1, 1, 7, 8, 1, 3, 3]

The appearance of the huge partial quotient 384859869703995923 in the continued fraction of zz is due, of course, to the fact that it is so close to a rational number of much smaller height, namely 691/638512875 (for a brief discussion of heights see §7.1). If we were given zz instead of z as an approximation to $\zeta(12)/\pi^{12}$ we would take as it mostly likely value the continued fraction truncated just before this large quotient. This is the same as running recognize(zz*1.)

```
? recognize(zz)
```

216440428080639720851361063/2000000000000000000000000000000000
```
? recognize(zz*1.)
```

691/638512875

We have been assuming, tacitly, that we can compute as good an approximation to our number as we want. In practice this may not be the case. If we know an a priori bound Q for the denominator of the number $r \in \mathbf{Q}$ that x approximates then a good bet for r is the convergent p_n/q_n, obtained from the continued fraction expansion of x, with the largest $q_n \leq Q$.

This is the *best approximation* to x with a given denominator bound Q and is what the GP function bestappr computes. Best approximation means the unique fraction p/q which minimizes

$$|qx - p|, \quad 1 \leq q \leq Q. \tag{1.12}$$

So for example,

```
? bestappr(.334,300)

1/3
```

We should stress that this, in general, is not the same as the fraction p/q which minimizes

$$|x - p/q|, \quad 1 \leq q \leq Q, \tag{1.13}$$

see below and Ex. 3. Also we may use bestappr for the following alternate version of recognize

```
recognize2(x)=bestappr(x, 10^(precision(x) - 5))
```

What do we do if we only know that r is rational but have no way to improve our approximation x of it? In a wonderful exercise D. Knuth, [52] vol. 2 §4.5.3 Ex. 39, asks at least how many times a baseball player has batted if his average is .334. (The batting average, is the rational number $0 \leq H/B \leq 1$ rounded to three decimal places, where H is the number of hits and B the number of battings.) To solve the problem we need to find the fraction $.3335 \leq p/q \leq .3345$ with smallest possible q. The answer is 96/287 (not 1/3!) and hence the player must have batted at least 287 times.

The solution to the general problem of finding $p/q \in [a, b]$ with the smallest possible q is described in [52] and item 101C by R. Gosper in [6]. Here is a GP implementation

```
gosper(v) =
{
    local(a,b,j,l,c);

    a=contfrac(v[1]);
    b=contfrac(v[2]);
    j=1; l=min(length(a),length(b));

    while(j <= l && a[j] == b[j],j++);

    c=vector(j-1,k,a[k]);
      if(j <= l,
        c=concat(c,if(a[j] < b[j],a[j],b[j])+1));
    c=contfracpnqn(c);
    c[1,1]/c[2,1]
}
```

We can now write a variant of `recognize` which will compute the fraction p/q in the interval $[x-\epsilon, x+\epsilon]$ with the smallest possible q, where $\epsilon = .5 \times 10^{-n}$ and n is either given or determined by the precision of x. WARNING: watch out if you input a rational number for x!

```
recognize1(x,n) =
{
    local(e);

    if(n==0,n=precision(x));
    e=.5/10^n;
    gosper([x-e,x+e])
}
```

For example,

```
? recognize1(0.334,3)

96/287
```

Here are the errors for the approximations $1/3$ and $96/287$ to 0.334 measured as in (1.12)

```
? \p 9
   realprecision = 9 significant digits
? 3*0.334-1

0.00199999986

? 287*0.334-96

-0.142000020
```

and as in (1.13)

```
? .334-1/3

0.000666666620

? .334-96/287

-0.000494773587
```

1.2.2 Rational numbers modulo *m*

Suppose that we are given integers $0 \leq c < m$ and we are told

$$\frac{a}{b} \equiv c \bmod m, \tag{1.14}$$

for some relatively prime integers a, b; can we recover a and b? This is indeed possible if $|a|, |b| \leq \sqrt{m/2}$ using continued fractions (see [62], Chapter 4, [97], [24]). Write $bc = a + dm$ for some $d \in \mathbf{Z}$. Then

$$\left| \frac{c}{m} - \frac{d}{b} \right| = \left| \frac{a}{bm} \right| \leq \frac{1}{2b^2}$$

and hence ([46] Theorem 184, p. 153) d/b is a convergent of c/m. Therefore, the following GP routine recovers a/b from c mod m.

```
recognizemod(c,m)=
{
  local(z,N);
  N=floor(sqrtint(floor(m/2)));
  z=bestappr(c/m,N);
  c-m*z
}
```

Note that we use `floor(sqrtint(floor(m/2)))` instead of `floor(sqrt(m/2))`. The reason for this is to avoid running into low precision problems when attempting to compute $\lfloor \sqrt{m/2} \rfloor$ for large m. We leave as an exercise (Ex. 5) to check that these two expressions are mathematically the same. In fact, since we will use this expression several times in what follows we write it as separate function

```
floorsqrt(x)=sqrtint(floor(x))
```

1.2.3 Algebraic numbers

If we are given a good (complex floating point) approximation to an algebraic number α of, say, degree 10, can we find its minimal polynomial? We can if the answer involves only reasonably small numbers.

The idea is to convert the question into one about finding short length vectors in a lattice and then apply lattice reduction algorithms (like LLL due to Lenstra, Lenstra and Lovasz see [60]) to solve that. This is what the GP function `algdep` does (based on the main routine `lindep` which finds integer linear dependences between given complex numbers). Concretely, assume for simplicity that α is real and consider the quadratic form

$$Q(a_0, a_1, \ldots, a_n) = C(a_n \alpha^n + \cdots + a_1 \alpha + a_0)^2 + a_0^2 + \cdots + a_n^2$$

for some (large) constant C. A small value of $Q(a_0, a_1, \ldots, a_n)$ with $a_i \in \mathbf{Z}$ means small a_i's with a small value of the polynomial $a_n x^n + \cdots a_1 x + a_0$ evaluated at α, which is precisely what we want.

Here is an example in GP.

```
? a=1+sqrt(1+sqrt(1+sqrt(2)))

2.5980531824786174203541125711
```

This is an approximation to $\alpha := 1 + \sqrt{1 + \sqrt{1 + \sqrt{2}}}$ an algebraic integer of degree at most 8. We try to find to find its minimal polynomial with

```
? algdep(a,8)

x^8 - 8*x^7 + 24*x^6 - 32*x^5 + 14*x^4 + 8*x^3 - 8*x^2 - 1
```

This looks like the correct answer as the coefficients are fairly small and do not change if we recalculate to a higher precision. In this case it is in fact easy to *prove* that it is correct (Ex. 7). In general, as with recognize, we can only make a good guess.

For example, we know π is transcendental; what happens if we try to a find a polynomial equation with integer coefficient for it?

```
? algdep(Pi,5)

 909*x^5 - 3060*x^4 + 1814*x^3 - 3389*x^2 - 723*x - 626
? \p 57
   realprecision = 57 significant digits

?  algdep(Pi,5)

16352996*x^5 - 41843628*x^4 - 17559998*x^3 - 25865436*x^2 -
49411721*x + 26594365

? subst(%,x,Pi)

8.549928927631901494 E-40
```

What we obtain is a fairly meaningless polynomial with large coefficients which has a very small value at $x = \pi$. It is a higher degree analogue of trying to recognize π as a rational number.

```
? \p 9
   realprecision = 9 significant digits
? recognize(Pi)

 104348/33215
? \p 28
   realprecision = 28 significant digits
? recognize(Pi)
```

```
    139755218526789/44485467702853
? \p 57
    realprecision = 57 significant digits
? recognize(Pi)

5368602728917869474170470223/17088793236078350364533888675
```

Remark A good approximation for our unknown number could also be *p*-adic. For example,

```
? a=sqrt(-1 + O(5^10))

 2 + 5 + 2*5^2 + 5^3 + 3*5^4 + 4*5^5 + 2*5^6 + 3*5^7 +
3*5^9 + O(5^10)

? algdep(a,2)

x^2 + 1
```

We end with a less trivial example. The $\eta(\tau)$ function of Dedekind is defined for τ in the upper half-plane by the infinite product

$$\eta(\tau) := e^{2\pi i \tau/24} \prod_{n=1}^{\infty}(1-q^n), \qquad q := e^{2\pi i \tau}, \qquad \Im(\tau) > 0. \qquad (1.15)$$

The theory of complex multiplication (from now on CM, see §3.2.5) tells us that certain ratios of values of η are algebraic (more precisely they are in abelian extension of imaginary quadratic fields). For example, Abel showed that

$$\frac{\eta^2(\sqrt{-5}/2)}{2\eta^2(2\sqrt{-5})} = \tfrac{1}{2}(1+\sqrt{5}) + \sqrt{\tfrac{1}{2}(1+\sqrt{5})}. \qquad (1.16)$$

Let us check it out.

```
? eta(sqrt(-5)/2,1)^2/2/eta(2*sqrt(-5),1)^2

 2.8900536382639638124570009296 + 0.E-28*I
? (1+sqrt(5))/2+sqrt((1+sqrt(5))/2)

 2.8900536382639638124570009296
? algdep(%,4)

x^4 - 2*x^3 - 2*x^2 - 2*x + 1
```

This calculation is not a proof of (1.16). We should point out, however, that given some a priori knowledge on the number we are computing (a bound on its degree and height) one can actually prove that the output polynomial of algdep is correct (see [7]).

It is of invaluable help to have these tools available when investigating certain questions. At the end of the day we may toss all calculations aside and keep only the proofs we found. But the proofs might not have been found without the intervening calculations.

In this situation of CM one can in principle figure out expressions for all the Galois conjugates $\alpha_1, \ldots, \alpha_n$ of a number like the left hand side of (1.16). We often also know in advance that the numbers are actually algebraic integers. In this case it is then easy to compute *exactly* what the minimal polynomial is without any guesswork. (Given approximations to $\alpha_1, \ldots, \alpha_n$ we just multiply out $\prod_i (x - \alpha_i)$; its coefficients are then approximations to rational integers.) But then again finding the Galois conjugates may not be that simple in actual practice.

Here are some more examples analogous to that of Abel. For negative discriminants $D \equiv 1 \bmod 8$ we define

$$u_D := \frac{1}{\sqrt{2}} \left| \frac{\eta(z_0/2)}{\eta(z_0)} \right|^2, \quad z_0 := \tfrac{1}{2}(1 + \sqrt{D}). \quad (1.17)$$

By CM theory u_D is a unit and $K(u_D)$ with $K = \mathbf{Q}(\sqrt{D})$ is an abelian extension (it is contained in the Hilbert class field of K, see §3.2.5, if D is coprime to 3; otherwise u_D^3 is).

In GP we may compute with the following script

```
ellunit(D)=
{
  local(z);
  z=(1+sqrt(D))/2;
  norm(eta(z/2,1)/eta(z,1))/sqrt(2)
}
```

Here are the minimal polynomial of the first few cases.

```
? for(d=8,100,if(d%8==7,u=ellunit(-d);
       if(d%3==0,u=u^3);h=qfbclassno(-d);
  print(-d,"\t",algdep(u,h))))

-15     x^2 - x - 1
-23     x^3 - x - 1
-31     x^3 - x^2 - 1
-39     x^4 - 3*x^3 - 4*x^2 - 2*x - 1
-47     x^5 - x^3 - 2*x^2 - 2*x - 1
```

```
-55        x^4 - 2*x^3 + x - 1
-63        x^4 - 8*x^3 + x + 1
-71        x^7 - 2*x^6 - x^5 + x^4 + x^3 + x^2 - x - 1
-79        x^5 - 3*x^4 + 2*x^3 - x^2 + x - 1
-87        x^6 - 13*x^5 - 11*x^4 + 4*x^3 - 4*x^2 - x - 1
-95        x^8 - 2*x^7 - 2*x^6 + x^5 + 2*x^4 - x^3 + x - 1
```

For details and more see for example [45]. There is a large literature on *elliptic units* such as u_D.

Technical remark To compute $\eta(\tau)$ it is more convenient to use its expression as a theta series (5.6) which converges very fast for $|q|$ small (large value of $\Im(\tau)$). We can exploit the fact that $\eta(\tau)$ satisfies the transformations formulas

$$\eta(\tau+1) = e^{2\pi i/24}\eta(\tau), \qquad \eta\left(\frac{-1}{\tau}\right) = \frac{1}{\sqrt{i\tau}}\eta(\tau)$$

to always reduce the calculation to evaluating $\eta(\tau_0)$ for τ_0 in the standard fundamental domain for the action of $SL_2(\mathbf{Z})$ on the upper half-plane (see §3.1.1). For such a τ_0 we have $\Im(\tau_0) \geq \sqrt{3}/2$ and hence $|q_0| \leq 0.0045$ where $q_0 = e^{2\pi i \tau_0}$. This is implemented in the GP has a built-in function eta(x,1).

1.3 Bernoulli polynomials

1.3.1 Definition and properties

Bernoulli numbers and polynomials $B_n(x)$ are pervasive in mathematics; they can be defined, for example, by means of the generating function

$$\frac{te^{tx}}{e^t - 1} = \sum_{n=0}^{\infty} B_n(x)\frac{t^n}{n!}, \quad B_n = B_n(0). \tag{1.18}$$

Here is a list of the first few

$$B_0(x) = 1$$
$$B_1(x) = x - \tfrac{1}{2}$$
$$B_2(x) = x^2 - x + \tfrac{1}{6}$$
$$B_3(x) = x^3 - \tfrac{3}{2}x^2 + \tfrac{1}{2}x$$
$$B_4(x) = x^4 - 2x^3 + x^2 - \tfrac{1}{30}.$$

The Bernoulli numbers and polynomials satisfy a veritable wealth of formulas and relations; we will need the following two.

$$B_n(x+1) - B_n(x) = nx^{n-1} \tag{1.19}$$

and

$$B'_n(x) = nB_{n-1}(x). \tag{1.20}$$

1.3.2 Calculation

If we want to find $B_n(x)$ for very small n, say to check a routine for computing them, we can always use the above definition directly, for example, as follows

```
? serlaplace(x*exp(t*x)/(exp(x)-1)+O(x^5))
1 + (t - 1/2)*x + (t^2 - t + 1/6)*x^2 +    (t^3 - 3/2*t^2 +
1/2*t)*x^3 + (t^4 - 2*t^3 + t^2 - 1/30)*x^4 +  O(x^5)
```

For larger values of n, however, we should do something else.

There can be quite a difference between the way that a computer and a human being view a formula. For example, let us consider the following two formulas for the Bernoulli polynomials:

$$B_n(x) = \sum_{k=1}^{n+1} \sum_{j=0}^{k-1} (-1)^{k+1} \frac{1}{k} \binom{n+1}{k} (x+j)^n \tag{1.21}$$

$$B_n(x) = \sum_{k=1}^{n+1} \sum_{j=0}^{k-1} (-1)^j \frac{1}{k} \binom{k-1}{j} (x+j)^n. \tag{1.22}$$

The first formula seems to be originally due to Kronecker [53]; it was rediscovered by D. Zagier from whom I learnt of it. See [40] for some history on both of these and other formulas; see §1.3.4 for a proof.

Either formula could be used to define $B_n(x)$ but a priori neither looks particularly simple or useful. On close inspection however, and despite both formulas looking quite similar, the first turns out to be very well suited for programming whereas the second is less so. The reason is that in the first, the terms in the sum over j are really independent of k and we should have written it as

$$B_n(x) = \sum_{k=1}^{n+1} (-1)^{k+1} \frac{1}{k} \binom{n+1}{k} \sum_{j=0}^{k-1} (x+j)^n \tag{1.23}$$

(we did not do it in order to stress the similarities of both formulas). Here a GP version coded by D. Zagier.

```
bernoulli(n,x)=
{
  local(h,s,c);
  h=0;s=0;c=-1;
    for(k=1,n+1,
      c*=1-(n+2)/k;
      s+=x^n;
      x++;
      h+=c*s/k);
  h
}
```

Let us analyze this function. The first thing to notice is that there is only one `for` loop while there are two summation signs in the formula (1.23). The point is that the summation over j in the formula is performed in the program one step at a time as k increases keeping the result in the variable s. Similarly, the binomial coefficient $\binom{n+1}{k}$ is computed incrementally with k and kept in the variable c (we have already seen this type of reasoning in §1.1.4). Note that the input x could be just about anything, as long as dividing by integers is not an issue (for p-adic inputs, however, see `bernoullip` in §6.3.3). If we want $B_n(x)$ as a *polynomial* we input the *variable* x; if we want a *value* of $B_n(x)$ at, say, $x = x_0$ we just input the *number* x_0. For example, the Bernoulli number B_{10} can be computed as follows

```
? bernoulli(10)

5/66
```

(note that there is no need to input 0 for x as GP automatically assigns 0 to any variable that it is not initialized explicitly).

A coded version of the other formula (1.22) would a priori require a double `for` loop and would hence be a lot slower.

1.3.3 Related questions

Differentiating (1.23) and using (1.20) we obtain recursively that

$$B_h(x) = \sum_{k=1}^{n+1}(-1)^{k+1}\frac{1}{k}\binom{n+1}{k}\sum_{j=0}^{k-1}(x+j)^h, \quad h = 0, 1, \ldots, n. \quad (1.24)$$

Hence if we replace $(x+j)^n$ in (1.23) by $f(x+j)$ where f is an arbitrary polynomial of degree at most n and define

$$F(x) := \sum_{k=1}^{n+1} \frac{(-1)^{k+1}}{k} \binom{n+1}{k} \sum_{j=0}^{k-1} f(x+j) \qquad (1.25)$$

then by (1.19)

$$F(x+1) - F(x) = f'(x). \qquad (1.26)$$

This follows from the fact that the formula is linear in f and the $B_h(x)$ for $h = 0, 1, \ldots, n$ are a basis of the vector space of all polynomials of degree at most n.

Consequently, we can easily modify our previous function to solve the antidifference equation

$$F(x+1) - F(x) = f(x) \qquad (1.27)$$

where f is a given polynomial (a solution always exists and is defined up to an additive constant just like for antiderivatives, (1.25) makes a particular choice).

```
antidiff(f)=
{
  local(x,h,s,y,c,n);
  x=variable(f);
  h=0;s=0;c=-1;y=x;

  n=poldegree(f);
  f=intformal(f);
    for(k=1,n+1,
      c*=1-(n+2)/k;
      s+=subst(f,x,y);
      y++;
      h+=c*s/k);
    h
}
```

Solving the antidifference equation is useful, for example, to compute sums of the form

$$\sum_{n=a}^{b} f(n) \qquad (1.28)$$

where f is a polynomial and $a, b \in \mathbf{Z}$. Indeed, if F is any solution to (1.27) then

$$\sum_{n=a}^{b} f(n) = \sum_{n=a}^{b} (F(n+1) - F(n)) = F(b+1) - F(a) \qquad (1.29)$$

1.3 Bernoulli polynomials

(it is in fact in this form, with $f(x) = x^h$, that J. Bernoulli encountered his polynomials in 1713, see (1.19)). Here is how we could do this in GP by modifying the routines above.

```
polsum(f,a,b) =
{
  local(h,s,c,n);
  h=0;s=0;c=-1;b++;
  n=poldegree(f);
  f=intformal(f);
  for(k=1,n+1,
      c*=1-(n+2)/k;
      s+=subst(f,x,b)-subst(f,x,a);
      a++;b++;
      h+=c*s/k);
  h
}
```

Naturally, for $b - a$ of size comparable to the degree of f we do not need this function and can simply do the sum directly.

We should point out that the limits a,b do not have to be constants; for example,

```
? f=polsum(x^3+x-1,0,x)

1/4*x^4 + 1/2*x^3 + 3/4*x^2 - 1/2*x - 1

? sum(n=0,100,n^3+n-1)-subst(f,x,100)

0
```

1.3.4 Proofs

There are many ways to present the theory of Bernoulli polynomials. We give here a brief sketch following the approach in [59]. Given $m \in \mathbb{N}$ we consider the following Q-linear operator acting on the space of polynomials $\mathbb{Q}[x]$.

$$T_m f(x) := \frac{1}{m} \sum_{k=0}^{m-1} f\left(\frac{x+k}{m}\right). \tag{1.30}$$

Fix an integer $m > 1$. It is clear that T_m preserves degrees and that in the basis $1, x, x^2, \ldots$ it is given by an upper triangular matrix with $1, m^{-1}, m^{-2}, \ldots$ along the diagonal. It follows that T_m has distinct rational eigenvalues $1, m^{-1}, m^{-2}, \ldots$ and hence there exists a Q-basis $A_0, A_1, A_2, \ldots$ of polynomials with $\deg(A_n) = n$ and $T_m A_n = m^{-n} A_n$. We can fix A_n uniquely by prescribing that they be monic. We will see shortly that A_n

is none other than the n-th Bernoulli polynomial, as defined for example in (1.18).

Lemma 1.2 *Fix an integer $m > 1$. Let E be a non-zero $\mathbf{Q}$-linear operator on $\mathbf{Q}[x]$ such that for some $e \in \mathbf{Q}^\times$*

$$T_m E = e E T_m. \tag{1.31}$$

Then $e = m^k$ for some $k \in \mathbf{Z}$ and for every $n \geq 0$ there exists a $\lambda_n \in \mathbf{Q}$ such that

$$EA_n = \lambda_n A_{n-k}, \quad n \geq k \tag{1.32}$$

and $EA_n = 0$ if $0 \leq n < k$.

Proof Since E is non-zero and the A_n form a basis of $\mathbf{Q}[x]$ there exists an n such that $EA_n \neq 0$. We have

$$T_m(EA_n) = eET_m A_n = m^{-n} e(EA_n).$$

Since EA_n is non-zero and the eigenvalues of T_m are all powers of m we must have $e = m^k$ for some $k \in \mathbf{Z}$. The same calculation for an arbitrary n shows that in general we have $T_m(EA_n) = m^{-n+k}(EA_n)$. Hence $EA_n = \lambda_n A_{n-k}$ for some $\lambda_n \in \mathbf{Q}$ if $n \geq k$ and $EA_n = 0$ if $0 \leq n < k$ as claimed. □

As a first example consider $E = D$ where $D = d/dx$ is derivation with respect to x in $\mathbf{Q}[x]$. By the chain rule we have for any $m \in \mathbf{N}$

$$T_m D = m D T_m$$

as it is easy to check. Hence D satisfies the hypothesis of Lemma 1.2 with $k = 1$. By comparing leading coefficients we see that

$$DA_n = nA_{n-1}, \quad n \geq 1$$

(compare with (1.20)).

We leave it to the reader as a nice exercise to verify the following

Proposition 1.3 *For all $l, m \geq 1$ we have*

$$T_m T_l = T_{ml} = T_l T_m. \tag{1.33}$$

Applying Lemma 1.2 to $E = T_l$ ($k = 0$ in this case) we find that

$$T_l A_n = l^{-n} A_n \tag{1.34}$$

so that the A_n's are a basis of common eigenfunctions for all $T_m, m \geq 1$.

It is convenient to define

$$T_{-1} f(x) := f(1-x), \quad T_{-m} := T_{-1} T_m. \tag{1.35}$$

It is straightforward to check that (1.33) still holds for all $l, m \in \mathbf{Z}$. Again by Lemma 1.2 with $E = T_{-1}$ we verify that

$$A_n(1-x) = (-1)^n A_n(x). \tag{1.36}$$

Finally we prove

Proposition 1.4 *For all $n \geq 0$ we have*

$$A_n = B_n.$$

Proof Since both A_n and B_n are monic it will be enough to show that B_n is an eigenvector for some T_m with $m > 1$. Fix an integer $m > 1$. We extend the operator T_m to power series in t with coefficients in $\mathbf{Q}[x]$. Applying it to both sides of (1.18) we obtain

$$\sum_{n=0}^{\infty} T_m B_n(x) \frac{t^n}{n!} = T_m \frac{te^{xt}}{e^t - 1}$$

$$= (e^t - 1)^{-1} \frac{t}{m} \sum_{k=0}^{m-1} e^{(x+k)t/m}$$

$$= \frac{\frac{t}{m} e^{x \frac{t}{m}}}{e^{\frac{t}{m}} - 1}$$

$$= \sum_{n=0}^{\infty} B_n(x) \frac{(t/m)^n}{n!}.$$

Hence $T_m B_n = m^{-n} B_n$ and we are done. $\square$

We may now give a proof of (1.23). For $n \in \mathbf{N}$ define the **Q**-linear operator on $\mathbf{Q}[x]$

$$F_n := \sum_{k=1}^{n+1} (-1)^{k+1} \frac{1}{k} \binom{n+1}{k} \sum_{j=0}^{k-1} T^j \tag{1.37}$$

where $Tf(x) := f(x+1)$. Then the more general statement (1.24) can be formulated as $F_n(x^h) = B_h(x)$ for $h = 0, 1, \ldots, n$.

The inner sum in (1.37) is a geometric series which equals $(T^k - 1)/(T - 1)$. On the other hand Taylor's theorem implies that

$$T = e^D := \sum_{l \geq 0} \frac{D^l}{l!} \tag{1.38}$$

(note that the series on the right actually becomes a finite sum when applied to any polynomial in Q[x] and hence defines a Q-linear operator in Q[x]).
It follows that

$$(T-1)F_n = \sum_{k=1}^{n+1}(-1)^{k+1}\frac{1}{k}\binom{n+1}{k}\sum_{l\geq 1}\frac{(kD)^l}{l!}$$

$$= \sum_{l\geq 1}\frac{D^l}{l!}\sum_{k=1}^{n+1}(-1)^{k+1}k^{l-1}\binom{n+1}{k}.$$

We claim that
$$(T-1)F_n \equiv D \bmod D^{n+2}. \tag{1.39}$$

By Taylor's theorem (1.38) we deduce

$$F_n \equiv \frac{D}{e^D-1} \bmod D^{n+1}. \tag{1.40}$$

This implies that for polynomials of degree at most n (i.e., in the kernel of D^{n+1}) we have $F_n = D/(e^D-1)$. We leave it to the reader to verify that for all $h = 0, 1, \ldots$ we have

$$\frac{D}{e^D-1}x^h = B_h(x), \tag{1.41}$$

which finishes the proof.

To prove the claim (1.39) we make the change variables $T = e^U$ so that $Td/dT = d/dU$. Then for $l = 0, 1, \ldots, n-1$

$$\sum_{k=0}^n (-1)^k k^l \binom{n}{k} = \left(T\frac{d}{dT}\right)^l (1-T)^n \bigg|_{T=1}$$

$$= \left(\frac{d}{dU}\right)^l (1-e^U)^n \bigg|_{U=0}$$

$$= 0$$

and now the claim follows easily.

We can prove (1.22) in a similar way. In this case the operator to consider is

$$\sum_{k=1}^{n+1}\sum_{j=0}^{k-1}(-1)^j\frac{1}{k}\binom{k-1}{j}T^j = \sum_{k=1}^{n+1}\frac{1}{k}(1-T)^{k-1}.$$

Modulo D^{n+1} the right hand side is $\log T/(T-1) = D/(e^D-1)$ and we again use (1.41) to finish the proof. $\square$

1.4 Sums of squares

A classical question in number theory is whether a given integer n is the sum of k squares, i.e., whether we can solve the equation

$$n = x_1^2 + \cdots + x_k^2, \quad x_1, \ldots, x_k \in \mathbb{Z}. \qquad (1.42)$$

Questions about solutions of a Diophantine equation like (1.42) can be formulated in many different ways with various degrees of difficulty. For example,

1. Is there a solution?
2. Are there finitely many solutions?
3. If there are finitely many solutions, can we count them? or at least estimate their number?
4. Can we find a solution explicitly?
5. Can we find all solutions explicitly?

We illustrate below some of the aspects of these questions for our equation (1.42). First, the number of solutions is clearly finite. We denote by $r_k(n)$ its number. A theorem of Lagrange [46], p. 302, guarantees that every positive integer n is the sum of four squares and one can characterize precisely those n which are sums of two squares (in terms of their prime factorization, see(1.4.4)) or three squares.

A priori, it is very easy to, say, list all solutions to (1.42) for a given n explicitly. Indeed, it is clear that for each i we have $|x_i| \leq \sqrt{n}$ and hence we can just run through all possible values of x_i and check whether they are solutions or not keeping those that are. This can be done in GP as follows.

```
sumsquares(n, k) =
{
  local(bd = sqrtint(n));

    forvec(x = vector(k,i, [-bd,bd]),
      if (n == sum(i=1, k, x[i]^2),
        print(x)))
}
```

(see the remark at the end of §1.4.4 for a description of the forvec construction used here). But note that (1.42) has a large group of symmetries, namely the group of order $2^k k!$ obtained by permuting and negating the coordinates, which we are not exploiting. For small n and k this is likely not to be relevant but it pays off eventually to be more careful in our programming.

Using the symmetries we can always take a solution to (1.42) to one satisfying
$$n = x_1^2 + \cdots + x_k^2, \quad 0 \leq x_1 \leq x_2 \leq \cdots \leq x_k \tag{1.43}$$
and our routines below will only find solutions of this form. Note, that the total number of solutions to (1.43) is not easily related in general to $r_k(n)$, which from many points of view is the right counting function to consider (for example it relates to modular forms §1.4.5).

1.4.1 $k = 1$

The question here is to know if a given positive integer n is a square or not. To do this we compute $m = \lfloor \sqrt{n} \rfloor$ analytically and compare n with m^2. If they agree n is a square (of m of course) otherwise n is not a square. GP has a built-in function issquare which does this (see below for more details).

1.4.2 $k = 2$

To find all solutions to (1.42) with $k = 2$ we can proceed as follows. First we check some obvious necessary conditions for there to be a non-trivial solution at all. Namely, we check that $n \geq 0$ and that $n \equiv 0, 1 \bmod 4$. (Such necessary congruence or sign conditions for an equation to have an integer solution are called *local* conditions.)

If n passes this test then we run over all integers $0 \leq x_1 \leq \sqrt{n/2}$ and check whether $n - x_1^2$ is a square or not. Here is how this procedure would look like in GP.

```
sum2squares(n) =
{
  local(x1, x2, v);

    if (n<0 || n%4 == 3, return([]));
    for (x1=0, floorsqrt(n/2),
       if (issquare(n - x1^2, &x2), v=concat(v,[[x1,x2]])));
    v
}
```

We should explain the use of issquare. As we said we want to test whether $n - x_1^2$ is a square. The output of issquare is 1 if it is and 0 otherwise; in the first case, however, GP has in the process computed a square root and this can be retrieved by the second argument to the function. More precisely, if x is a non-negative integer then

```
issquare(x^2,&y)
```

outputs 1 (x^2 is indeed a square) and sets the value of the variable y to x.

1.4 Sums of squares

If the input is a sum of two squares then the output is a vector whose components are two-component vectors $[x_1, x_2]$ with all pairs satisfying $0 \le x_1 \le x_2$ and $x_1^2 + x_2^2 = n$; otherwise, the output is empty.

If we only want to know whether there is a solution or just find one when it exists we can improve the average running time of our function by getting out of the loop once we found a solution. This can be done in GP as follows.

```
sum2squares1(n) =
{
  local(x1, x2);

    if (n<0 || n%4 == 3, return([]));
    for (x1=0, floorsqrt(n/2),
       if (issquare(n - x1^2, &x2), return([x1,x2])));
}
```

We need to initialize x_1 as -1 since until *does first and asks later*; in our case, x_1 is actually changed to 0 before any testing is done. This should be contrasted with while, which *asks first and does later*. The difference between until and while can be confusing.

We come out of the until loop when either $x_1 > $ bd, in which case we have ran through all possible values of x_1 without finding a solution, or $n - x_1^2$ is a square, in which case we have a solution.

1.4.3 $k \ge 3$

Much of what we did for the case of sums of two squares can be generalized to any k. As an illustration we give the analogue of sum2squares1 for $k = 3$ and 4 (by Lagrange's theorem we need to go no further for this function). Here we avoided until or while entirely in favor of the simpler to use return, which takes us out of the loop whenever we want to.

For $k = 3$ we have

```
sum3squares1(n) =
{
  local(x_1,x_2,x_3,n_1);

  if(n<0 || n%8==7,return([]));
       for(x_1=0,floorsqrt(n/3),
           n_1=n-x_1^2;
       for(x_2=x_1,floorsqrt(n_1/2),
           if(issquare(n_1-x_2^2,&x_3),
               return([x_1,x_2,x_3]))))
}
```

Note the local obstruction at 2: a number $\equiv 7 \bmod 8$ cannot be a sum of 3 squares.

For $k = 4$

```
sum4squares1(n) =
{
  local(x_1,x_2,x_3,x_4,n_1,n_2);

  if(n<0,return([]));
        for(x_1=0,floorsqrt(n/4),
            n_1=n-x_1^2;
         for(x_2=x_1,floorsqrt(n_1/3),
            n_2=n_1-x_2^2;
          for(x_3=x_2,floorsqrt(n_2/2),
            if(issquare(n_2-x_3^2,&x_4),
              return([x_1,x_2,x_3,x_4])))))
}
```

Note that there are no local conditions at 2 for $k = 4$.

Here is a list of the output of these functions for $1 \leq n \leq 20$.

```
? for(n=1,20,print(n,"\t",sum3squares1(n)))

1         [0, 0, 1]
2         [0, 1, 1]
3         [1, 1, 1]
4         [0, 0, 2]
5         [0, 1, 2]
6         [1, 1, 2]
7         []
8         [0, 2, 2]
9         [0, 0, 3]
10        [0, 1, 3]
11        [1, 1, 3]
12        [2, 2, 2]
13        [0, 2, 3]
14        [1, 2, 3]
15        []
16        [0, 0, 4]
17        [0, 1, 4]
18        [0, 3, 3]
19        [1, 3, 3]
20        [0, 2, 4]

? for(n=1,20,print(n,"\t",sum4squares1(n)))

1         [0, 0, 0, 1]
2         [0, 0, 1, 1]
```

3	[0, 1, 1, 1]
4	[0, 0, 0, 2]
5	[0, 0, 1, 2]
6	[0, 1, 1, 2]
7	[1, 1, 1, 2]
8	[0, 0, 2, 2]
9	[0, 0, 0, 3]
10	[0, 0, 1, 3]
11	[0, 1, 1, 3]
12	[0, 2, 2, 2]
13	[0, 0, 2, 3]
14	[0, 1, 2, 3]
15	[1, 1, 2, 3]
16	[0, 0, 0, 4]
17	[0, 0, 1, 4]
18	[0, 0, 3, 3]
19	[0, 1, 3, 3]
20	[0, 0, 2, 4]

Remark Instead of writing sum3squares1 from scratch we could also use sum2squares1 for the innermost loop (and similarly sum4squares1 could use sum3squares1).

1.4.4 The numbers $r_k(n)$

A fair amount is known about $r_k(n)$, defined in §1.4, especially for small k. For example, for $k=2$ and $k=4$ we have the following explicit formulas [46], p. 314 (for $n > 0$)

$$r_2(n) = 4\sum_{d|n} \chi(d) \tag{1.44}$$
$$r_4(n) = 8\sum_{d|n} \delta(d)d \tag{1.45}$$

where χ is the primitive quadratic character of conductor 4 and

$$\delta(d) = \begin{cases} 1 & \text{if } 4 \nmid d \\ 0 & \text{if } 4 \mid d \end{cases}$$

(In particular, $r_4(n) > 0$ for $n > 0$ since $\delta(1) = 1$ and all terms in the sum are non-negative, which implies Lagrange's theorem.) Note, however, that computationally these formulas will not be that useful for very large n as they require we know a factorization of n. In any case, in GP they are very easy to program thanks to the built-in function sumdiv(n,d,expr) which gives the sum of the expression expr over all divisors d of n:

sum2squares2(n)=4*sumdiv(n,d,kronecker(-4,d))

sum4squares2(n)=8*sumdiv(n,d,if(d%4,d))

For comparison, we can easily modify our previous script `sum2squares` so that it returns the total number of solutions as follows

```
sum2squares3(n) =
{
  local(x_1,x_2,v);

  ct=0;

  if(n<0 || n%4==3,return(0));
    for(x_1=0,floorsqrt(n/2),
    if(issquare(n-x_1^2,&x_2),
      ct+=((x_1>0)+1)*((x_2>0)+1)*((x_1<x_2)+1)));
  ct
}
```

Note again the mixture of boolean and standard arithmetic: the conditionals `(x_1>0)`, `(x_2>0)`, and `(x_1<x_2)` have the value 1 or 0 according to whether the corresponding statements are valid or not; GP has no problem then using that answer in standard arithmetic statements like `(x_1>0)+1`, etc.

Here is a cleaner and faster version (due to K. Belabas)

```
sum2squares4(n) =
{
  local(ct);

    if (n < 0 || n%4 == 3, return (0));
    if (n == 0, return (1));
    ct = 4 * (issquare(n) + issquare(n/2));
    for (x1 = 1, sqrt((n-1)/2),
      if (issquare(n-x1^2), ct += 8));
    ct
}
```

If $n = p$ is an odd prime then $r_2(p) = 8$ if $p \equiv 1 \bmod 4$ and is 0 otherwise; in other words, every prime congruent to 1 modulo 4 is a sum of two squares in an essentially unique way. This theorem was proved by Fermat; see [99] for a very short proof.

If

$$p = x^2 + y^2, \quad x, y \in \mathbb{Z} \qquad (1.46)$$

then $(x/y)^2 \equiv -1 \bmod p$ (by size considerations $p \nmid y$). Conversely, starting with a square root z of $-1 \bmod p$, i.e., $z^2 \equiv -1 \bmod p$, we can find a solution to (1.46) using Gauss reduction of binary quadratic forms. This gives a very efficient algorithm to solve (1.46) due to Cornachia see §3.1.2. Using the GP built-in functions this can be implemented as follows.

```
fermat(n)=qfbsolve(Qfb(1,0,1),n)
```

As an example we have

```
? fermat(10^50+577)
```

```
[-648626890687392164245424, 761106534380835424545450401]
```

For a non-zero integer n and a prime p we define $v_p(n)$, the *valuation* of n at p, as the exact power of p dividing n. One consequence of (1.44) is that for $n > 0$ we have

$$n = \square + \square \iff v_p(n) \text{ is even} \quad \text{for all } p \equiv 3 \bmod 4. \tag{1.47}$$

Remark A surprising consequence of the formulas (1.44) and (1.45) is that the arithmetic functions $\frac{1}{4}r_2(n)$ and $\frac{1}{8}r_4(n)$ are multiplicative. A function $f : \mathbf{N} \longrightarrow \mathbf{C}$ is *multiplicative* if $f(mn) = f(m)f(n)$ for any two coprime integers m and n. This reflects the fact that the equations (1.42) for $k = 2$ and $k = 4$ are related to the norm in a certain ring ($\mathbf{Z}[i]$ and the Hurwitz quaternions respectively, Ex. 17).

For $k = 3$ there is a relation between $r_3(n)$ and the class number of orders in imaginary quadratic fields (see Ex. 18), illustrating the general phenomenon that the behavior of $r_k(n)$ is more involved for odd values of k. Formulas for $r_k(n)$ do exist for other small values of k but are more complicated and for larger values there really is no simple closed formula. see [50], Chapter 11.

It is quite easy to write a program that will compute $r_k(n)$ for any k and n recursively by noticing that

$$r_k(n) = \sum_{x \in \mathbf{Z}} r_{k-1}(n - x^2), \quad k > 2. \tag{1.48}$$

In GP we could do it as follows (we assume that the input n is positive)

```
sumsq(n,k) =
{
  if(n<0,0,
    if(k==1,if(n,2*issquare(n),1),
      sumsq(n,k-1)+2*sum(x=1,sqrtint(n),sumsq(n-x^2,k-1))))
}
```

In general, however, recursive definitions, although often easier to write, tend to be slower than their non-recursive counterparts (in our case we could speed it up by stopping the recursion at $k = 2$ or 3). Moreover, our sumsq does not exploit all the symmetries of the problem. For example, for $k = 5$ it would be better to use the following

```
sum5squares1(n)=
{
  local(x_1,x_2,x_3,x_4,x_5,n_1,n_2,n_3,x,ct);

  ct=0;

    for(x_1=0,floorsqrt(n/5),
      n_1=n-x_1^2;

       for(x_2=x_1,floorsqrt(n_1/4),
         n_2=n_1-x_2^2;

          for(x_3=x_2,floorsqrt(n_2/3),
            n_3=n_2-x_3^2;

             for(x_4=x_3,floorsqrt(n_3/2),

               if(issquare(n_3-x_4^2,&x_5),
                 x=[x_1,x_2,x_3,x_4,x_5];
                 ct=ct+signsno(x)*permno(x))))));
  ct
}
```

where given a vector x

```
permno(x, sortfirst = 0)=
{
  local(w,a);
  if(sortfirst,x=vecsort(abs(x)));
  w=1; a=1; b=1;

  for(j=1,length(x)-1,
    w=w*(j+1);

    if(x[j]==x[j+1],
      b=b+1;a=b*a,
      w=w/a;a=1;b=1));
  w/a
}
```

returns the number of distinct permutations of x (given by $n!/\prod_i n_i!$, where each n_i is the size of a set of equal coordinates in x) and

```
signsno(v,sortfirst = 0)=
{
    local(j);

    if(sortfirst,v=vecsort(abs(v)));
```

```
        j=1;
        while(v[j]==0 && j<=length(v),j++);
        shift(1,length(v)-j+1)
}
```

returns 2^k where k is the number of non-zero entries in x. (In sum5squares1 we we could have simply used prod(j=1,5, (0<x[j])+1) instead of signsno but it would be formally less appealing.)

Note the optional variable sortfirst which if given a non-zero value will instruct the routines to sort the (absolute values of the coordinates of the) input vector. In our case the coordinates of the input are already sorted and non-negative so we skip this step by passing no value to sortfirst (hence set to zero by GP). We could also easily compute the total contribution

signsno(x)*permno(x)

in one go (Ex. 21).

The final routine is certainly longer and somewhat trickier to code than sumsq(n,5). Is it faster? We leave the analysis of these routines as an exercise (Ex. 19).

Remark It does not seem possible to write in GP a version of sum5squares1 that will take k as a variable; the issue is how to generate dynamically the series of nested for loops that we want. However, it is possible and quite convenient to write a series of nested loops with forvec. Say we want to run over all integer points in a box of the form $[a_1, b_1] \times \cdots \times [a_k, b_k]$ where the intervals $[a_i, b_i]$ for $i = 1, 2, \ldots, k$ are part of the input. We can achieve this simply as follows (assuming we already have the set value $v = [[a_1, b_1], \ldots, [a_k, b_k]]$)

```
    ...
    forvec(u=v,
        ...
    )
```

Then $u[i]$ will run through all integers in $[a_i, b_i]$.

If all the intervals are equal to $[a, b]$ then we just need a, b and k as inputs and we could write the routine as

```
    forvec(u=vector(k,j,[a,b]),
        ....
    )
```

The GP function also has an optional argument to run only over increasing (or strictly increasing) sequences.

1.4.5 Theta functions

There is yet another and much better way we can compute $r_k(n)$ (though not the actual solutions to (1.42)) using generating functions. If we let

$$\theta = \sum_{n \in \mathbb{Z}} q^{n^2} \qquad (1.49)$$

be a formal power series in the variable q then

$$\theta^k = \sum_{n=0}^{\infty} r_k(n) \, q^n. \qquad (1.50)$$

Hence if we wanted to compute *all* values of $r_k(n)$ for $n = 0, 1, \ldots, N$ we could do it in GP as follows. To fix ideas say $N = 100$ and $k = 5$. First we define θ to precision N

```
th = sum(n=1, 10, 2*q^(n^2),1+O(q^101));
```

(we use th as the name theta is reserved in GP for a built-in function) and then compute

```
th_5=th^5;
```

(we add the semicolon ; at the end of the line since we do not particularly care to see the whole series all at once). Now we may retrieve $r_5(n)$ for $n = 0, 1, \ldots, 100$ with

```
polcoeff(th_5,n)
```

If anything else this approach is very safe and can be used to check the correctness of a script computing $r_k(n)$ (certainly this is what we did while writing this section!). For large scale computations one can use techniques like the fast Fourier transform to compute θ^k.

The functions θ^k are actually a lot more than placeholders for the numbers $r_k(n)$ since they are in fact modular forms (see [50]).

1.5 Exercises

1. Write a GP routine that given a non-zero integer D as input returns the discriminant of the associated étale algebra $\mathbb{Q}(\sqrt{D})$.
2. Consider the quadratic form $Q(x,y) := 3x^2 + 2xy + 4y^2$. For p prime define $S(p) := N(p) - 1$, where $N(p)$ is the number of projective solutions to the congruence $Q(x,y) \equiv 0 \bmod p$ (projective means non-zero pairs (x,y) up to scaling). Write a GP routine to compute $S(p)$ and check its modularity numerically.
3. Write a GP routine that given x and Q computes p/q minimizing (1.13).

4. If the baseball player of Knuth's problem mentioned in §1.2.1 actually batted more than 287 times, at least how many times was that?
5. Prove that for all $x \geq 0$ we have $\lfloor \sqrt{\lfloor x \rfloor} \rfloor = \lfloor \sqrt{x} \rfloor$.
6. What is the minimal polynomial of $\sqrt[16]{-4}$?
7. Find the minimal polynomial of

$$1 + \sqrt{1 + \sqrt{1 + \sqrt{1 + \cdots \sqrt{2}}}}.$$

8. Prove that with appropriate choice of square roots

$$\sqrt{5} + \sqrt{22 + 2\sqrt{5}} = \sqrt{11 + 2\sqrt{29}} + \sqrt{16 - 2\sqrt{29} + 2\sqrt{55 - 10\sqrt{29}}}$$

9. Prove that
$$B_{2n} \sim (-1)^{n-1} 4\sqrt{\pi n} \left(\frac{n}{\pi e}\right)^{2n}, \quad n \to \infty \qquad (1.51)$$

and test it numerically.
10. Rewrite sum2squares1 using while instead of until.
11. Write a script to count the number of solutions to

$$x_1^2 + \cdots x_k^2 \equiv 1 \bmod p$$

for $k = 2, 3, \ldots$ up to, say, $k = 5$ (you can use forvec instead of a nested series of for loops) and make a guess for what the general formula is.
12. Using the script sgn, experiment with various polynomials to check Theorem 1.1.
13. Consider the function of a prime p given by

count1(p) =p-1-sum(x=0,p-1,kronecker(x^4-1,p)+1)

Describe what count1(p) has to do with the number of solutions in x and y of

$$y^2 \equiv x^4 - 1 \bmod p.$$

Make an educated guess of the relation between the values of count1 and fermat. See if you can do the same for the equation

$$y^2 \equiv x^4 - 25 \bmod p$$

and (harder)

$$y^2 \equiv x^4 - 5 \bmod p.$$

14. Test numerically the formulas (1.44) and (1.45) and convince yourself that they are correct.
15. Check that sum2squares3 and sum2squares4 work and compare their running times.
16. Modify sum2squares4 so that it gives a solution to

$$n = x^2 + y^2, \quad \gcd(x, y) = 1$$

when there is one (when is there one?).
17. Write a GP routine that tests whether a function $f(n)$ is multiplicative (up to a constant factor). Check to see which of the functions $r_k(n)$ for $k = 1, 2, \ldots, 10$

appear to be multiplicative (up to a constant factor). Research how your result is connected to the norm on a certain ring alluded to in §1.4.4.
18. Using sumsq (or your own script) to compute $r_3(n)$, together with the GP built-in function qfbclassno to compute class numbers of imaginary quadratic fields, guess by experimentation the formula (due to Gauss) relating both functions (at least for special n).
19. Analyze sum5squares1, permno, signsno and verify that they compute what is intended. Compare the running times of sum5squares1 and sumsq(n, 5) (with and without the modification to break the recursion at $k = 2$ or 3).
20. Check the validity of the following version of permno (due to K. Belabas)

```
permno1(x, n) =
{
  local(ct = vector(n));
    for (i=1, #x, ct[ x[i] ]++);
    n! / prod(i=1, n, ct[i]!)
}
```

and compare its running time with permno.
21. Write a function that computes signsno(x)*permno(x) in one go.
22. Compare the running times of sumsquares(n) (with $k = 5$) sum5squares1(n), and sumsq(n,5) for $n = 100, 200, \ldots$, etc. (Start the timer in GP with the command #.)
23. What is wrong with the following version of signsno?

```
signsnox(x) =
{
  local(j);

  x=vecsort(x);
  j=1;
    while(x[j]==0 && j<=length(x), j++);
  shift(1,length(x)-j+1)
}
```

24. Assuming the known fact that
$$r_5(n^2) = 10 \prod_p r'_5\left(p^{v_p(n)}\right)$$
where
$$(p^3 - 1)\, r'_5\left(p^{2k}\right)$$
is a polynomial in p, experiment with sumsquares to make a guess of a formula for $r_5(n^2)$. (The formula for $r'_5(p^{2k})$ has the same general shape for all $p > 2$.)
25. Why would

polcoeff(th^5,n)

as opposed to

polcoeff(th_5,n)

not be such a great idea?

26. Consider the arithmetic function

$$R_k(n) := \#\{x_1^2 + \cdots + x_k^2 \leq n, \quad x_1, \ldots, x_k \in \mathbb{Z}\} = \sum_{m=0}^{n} r_k(m),$$

which counts the number of integral points in a sphere of radius $\sqrt{n}$ in k-dimensional space. Show that $R_k(n)$ approximates the volume of that sphere for large n. Write a GP script to test this fact.

27. Write a script to compute the number of solutions to

$$n = x^3 + y^3, \quad x, y \in \mathbb{Z}_{\geq 0}$$

and verify Ramanujan's famous taxicab number claim (1729 is the smallest positive integer that can be written as a sum of two positive cubes in two different ways).

28. (Hard) Find the smallest cube-free integer which can be expressed as the sum of two positive cubes in three different ways.

2 Reciprocity

2.1 More variation with p

We saw in §2.1 how the quadratic reciprocity law can be interpreted as describing the variation of the number of solutions $N(p)$ of a quadratic equation modulo primes p. We will now consider the situation for polynomials still in one variable but of higher degree.

2.1.1 Local zeta functions

Let us fix a prime p. In order to formulate the general reciprocity law (of Artin) it turns out to be necessary to consider not just the number of solutions to an equation modulo p but the result of other related countings as well.

Given a system of polynomial equations

$$X : \begin{cases} F_1(x_1,\ldots,x_N) = 0 \\ \vdots \\ F_N(x_1,\ldots,x_N) = 0 \end{cases} \quad (2.1)$$

with coefficients in the finite field $\mathbf{F}_p$ we let

$$N_r = \#X(\mathbf{F}_{p^r}), \quad r \in \mathbf{N}, \quad (2.2)$$

where $\mathbf{F}_{p^r}$ denotes the field of p^r elements in a fixed algebraic closure of $\mathbf{F}_p$, be the number of solutions of the system (2.1) with $x_i \in \mathbf{F}_{p^r}$ for $i = 1,\ldots,n$. (Note that N_1 is precisely what we denoted $N(p)$ in §1.1.)

If X consists of only one equation $f(x) = 0$ with $f \in \mathbf{F}_p[x]$ of non-zero discriminant we can compute N_r as follows. Let

$$f = f_1 \cdots f_m$$

be a factorization of f into (distinct) irreducible factors $f_j \in \mathbf{F}_p[x]$ of degree n_j. Then

$$N_r = N_r^{(1)} + \cdots + N_r^{(m)} \quad (2.3)$$

where $N_r^{(j)}$ is the corresponding counting function for the system X_j defined by $f_j = 0$. This follows from two facts: (i) $\mathbb{F}_{p^r}$ is a field and hence a product is zero if and only if at least one factor is zero; and (ii) the f_i's are *distinct* irreducibles and hence share no root. On the other hand, if f is irreducible of degree n then it is not hard to see that

$$N_r = \begin{cases} n & \text{if } n \mid r \\ 0 & \text{if } n \nmid r \end{cases} \tag{2.4}$$

(If $\alpha \in \mathbb{F}_{p^r}$ is a root of f then $\mathbb{F}_p(\alpha) \subseteq \mathbb{F}_{p^r}$ so $n = [\mathbb{F}_p(\alpha) : \mathbb{F}_p] \mid r$. Conversely, if $n \mid r$ then $\mathbb{F}_{p^n} \subseteq \mathbb{F}_{p^r}$ and $\mathbb{F}_p[x]/(f)$ is isomorphic to $\mathbb{F}_{p^n}$ since they have the same cardinality.)

A convenient way to encode the information contained in (2.3) and (2.4) is to form the *(local) zeta function* of X

$$Z(X, T) = \exp\left(\sum_{r=1}^{\infty} \frac{N_r}{r} T^r\right) \tag{2.5}$$

where the expressions are treated as formal power series in the variable T.

Proposition 2.1 *Let X be given by the equation*

$$X : f(x) = 0$$

where $f \in \mathbb{F}_p[x]$ has non-zero discriminant. Then

$$Z(X, T) = \frac{1}{(1 - T^{n_1}) \cdots (1 - T^{n_k})} \tag{2.6}$$

if and only if f factors as

$$f = f_1 \cdots f_k$$

with $f_j \in \mathbb{F}_p[x]$ irreducible of degree n_j for $j = 1, \ldots, k$.

Let us order the degrees of the factors in descending order $n_1 \geq \cdots \geq n_k$ and call $\tau = [n_1, \ldots, n_k]$ the *factorization type* of f. Note that $\deg(f) = n_1 + n_2 + \cdots + n_k$ so τ is a *partition* of $\deg(f)$.

In conclusion, the numbers of solutions N_r are completely determined by the factorization type τ.

2.1.2 Formulation of reciprocity

Recall that a (finite) étale algebra A over $\mathbb{Q}$ is a $\mathbb{Q}$-algebra of the form

$$A = \mathbb{Q}[x]/(f),$$

where $f \in \mathbb{Q}[x]$ is a polynomial with non-zero discriminant. Let n be the degree of f.

For all but a finite set S_f of primes p the reduction of f modulo p makes sense (i.e., p does not divide the denominator of the coefficients of f) and has degree n and non-zero discriminant. For $p \notin S_f$, let τ_p be the factorization type of f modulo p. Note that the set S_f of excluded primes does depend on the choice of f and not just on the isomorphism class of A. However, the factorization type, when defined, only depends on the isomorphism class of A; i.e., τ_p is the same for all polynomials g with $A \simeq \mathbf{Q}[x]/(g)$ and $p \notin S_g$.

As opposed to the quadratic case a typical algebra does not have a global minimal model (we could mimic the definition we gave in the quadratic case but this notion is not really meaningful). However, one can define disc(A) in an intrinsic way (we omit its description as it would take us too far afield); disc(A) contains all the primes which behave in a special way for A (the *bad primes* for A or, technically, the primes that *ramify* in A). We will denote this (finite) set of primes by S_A. A prime not in S_A is called *unramified*. The partition τ_p of $[A : \mathbf{Q}]$ is in fact defined for any $p \notin S_A$ (though not in general as the factorization type at p of some polynomial equation for A).

For any polynomial $f \in \mathbf{Z}[x]$ such that $A \simeq \mathbf{Q}[x]/(f)$ we have disc(f) = m^2 disc(A) for some $m \in \mathbf{Q}^\times$ (in fact $m \in \mathbf{Z}$ if f is monic) but there just may not be any such polynomial with disc(f) = disc(A). (An important family of cases where there *is* such a polynomial equation, namely the cyclotomic polynomial, are the cyclotomic fields.)

If $A = \prod_{i=1}^r K_i$ with K_i a number field then disc(A) = $\prod_i$ disc(K_i). In GP we can compute disc(K) for a number field $K = \mathbf{Q}[x]/(f)$ with `nfdisc(f)` (functions whose name starts with `nf` pertain to number fields). In fact `nfdisc` computes the discriminant of an étale algebra (not necessarily a field) defined by its polynomial argument. WARNING: if the argument of `nfdisc` is not squarefree the function will give an error.

It will be useful to introduce the following terminology, analogous to that used in §1.1.2. We will say that A (or the map $p \mapsto \tau_p$) is *modular* if there exists a number $N \in \mathbf{N}$ such that, whenever defined, τ_p depends only on p modulo N. We will call the smallest such N the *conductor* cond(A) of the algebra A.

In this setting, we may reformulate the quadratic reciprocity law, in a weaker form, as follows.

Theorem 2.2 *Every quadratic étale algebra over $\mathbf{Q}$ is modular.*

(This version is weaker than the full reciprocity law since we do not specify N or the actual dependence of τ_p on p mod N.)

Artin's reciprocity law for **Q**, again in a weak form, is a similar statement for abelian étale algebras of higher degree. Recall that

$$A = \prod_{i=1}^{r} K_i,$$

where each $K_i = \mathbf{Q}[x]/(f_i)$ is a number field. We will say that A is *abelian* if each extension $K_i/\mathbf{Q}$ is abelian, i.e., if it is Galois and its Galois group is abelian.

Theorem 2.3 *A finite étale algebra over* **Q** *is modular if and only it is abelian.*

Clearly this theorem contains the previous one as any field extension $K/\mathbf{Q}$ of degree at most 2 is necessarily abelian. In the quadratic case the conductor and discriminant coincide; in general, what is true is that they have exactly the same prime factors.

Remark The full story of what is going on here is given by class field theory (see for example [26], Chapter II §8). In particular we have the Kronecker–Weber theorem: any finite abelian field extension $K/\mathbf{Q}$ is contained in a cyclotomic one $\mathbf{Q}(\zeta_m)/\mathbf{Q}$ (ζ_m a primitive m-th root of unity). The least such m is the conductor of K; the way K sits in $\mathbf{Q}(\zeta_m)$ gives a precise description of how τ_p depends on $p \bmod m$. See below, §2.2.2 for a simple example of this.

2.1.3 Global zeta functions

Given a finite étale algebra A there is a way to package the information of all factorization types τ_p in one single global object $\zeta_A(s)$ called the *zeta function* of the algebra. The zeta function is an analytic function of the complex variable s of the general form

$$\zeta_A(s) = \prod_p F_p(p^{-s})^{-1}, \quad \Re(s) > 1, \tag{2.7}$$

for certain polynomials $F_p(T)$ (called the *Euler factor at p* of ζ_A). As indicated this infinite product converges in the half-plane $\Re(s) > 1$. For an unramified prime p the Euler factor $F_p(T)$ is the denominator of the local zeta function; i.e., $F_p(T) = \prod_i (1 - T^{n_i})$ where $\tau_p = [n_1, n_2, \ldots]$ is the factorization type of p. In particular, it contains no more information than the type τ_p itself. For ramified primes F_p has a more involved definition which we will not give explicitly. The function ζ_A uniquely determines the

Euler factors. To describe all factorization types τ_p is then tantamount to describing the zeta function.

It is a fundamental fact of algebraic number theory that $\zeta_A(s)$ originally only defined for $\Re(s) > 1$ actually extends to a holomorphic function of s except for a simple pole at $s = 1$ and satisfies a functional equation when s is replaced with $1 - s$.

The quintessential example is of course the Riemann zeta function

$$\zeta(s) = \prod_p (1 - p^{-s})^{-1} = \sum_{n \geq 1} n^{-s}. \tag{2.8}$$

which corresponds to $A = \mathbf{Q}$.

2.2 The cubic case

Let us consider in some detail the case of irreducible algebras of degree 3; i.e., cubic extensions $K/\mathbf{Q}$. There are two possibilities: either $K/\mathbf{Q}$ is Galois, with $\text{Gal}(K/\mathbf{Q}) \cong \mathbf{Z}/3\mathbf{Z}$, or $K/\mathbf{Q}$ is not Galois and its Galois closure has Galois group isomorphic to the non-abelian group S_3. The two cases correspond to whether $\text{disc}(K)$ is a square or not, respectively. In general, $\text{disc}(K)$ is a square if and only if the Galois group of the Galois closure is, as a permutation group of the roots of f, a subgroup of A_n. (This should be compared with Theorem 1.1: $D = \text{disc}(K)$ is a square if and only if the corresponding Dirichlet character is trivial; by 1.1 this means that the sign of Frob$_p$, see §2.3, is always $+1$ and hence all Galois elements are even permutations of the roots.)

It is quite easy to write a GP function that will return the factorization type of a polynomial, for example

```
polfacttype(f,p) =
{
  if(!(poldisc(f)%p),return([]));
  vecsort(factormod(f,p,1)[,1]~,,4)
}
```

The built-in function factormod with the flag argument 1 does the job, the rest is cosmetic; the tagged on [,1] picks the first row of the result, which contains what we want: the degrees of the irreducible factors. If p divides the discriminant of f then τ_p is not defined and we output the empty vector.

2.2.1 Two examples

To illustrate the two cases, let us consider $f := x^3 + x^2 - 2x - 1$ and $g := x^3 + x + 1$. Their discriminants are 49 and -31, respectively, hence, $\mathbb{Q}[x]/(f)$ is abelian and $\mathbb{Q}[x]/(g)$ is not. We use the following GP function.

```
poltypes(f, a = 2, b = 1000, w = []) =
{
  forprime(p=a, b,
    v = polfacttype(f, p);
      if (v && !memb(v,w),w=concat(w,[v])));
  w
}
```

which outputs a list of the distinct non-trivial types appearing in the search performed (primes p from a to b with default values of 2 and 1000, respectively). The optional argument w is a precomputed list of types in case we want to extend a previously executed search.

We obtain

```
? poltypes(x^3 + x^2 - 2*x - 1)
```

```
[[3], [1, 1, 1]]
```

and for g

```
? poltypes(x^3 + x+ 1)
```

```
[[3], [2, 1], [1, 1, 1]]
```

If we actually want to see how the types depend on p we can use this function instead

```
pollisttype(f, bd = 50) =
{
  local(v);

  forprime(p=2,bd,
    v=polfacttype(f,p);
      if(!v, next);
      print(p,"\t",v))
}
```

Here is the corresponding output for the two polynomials.

```
? pollisttype(x^3 + x^2 - 2*x - 1)
```

```
2       [3]
3       [3]
5       [3]
```

11	[3]
13	[1, 1, 1]
17	[3]
19	[3]
23	[3]
29	[1, 1, 1]
31	[3]
37	[3]
41	[1, 1, 1]
43	[1, 1, 1]
47	[3]

```
? pollisttype(x^3+x+1)
```

2	[3]
3	[2, 1]
5	[3]
7	[3]
11	[2, 1]
13	[2, 1]
17	[2, 1]
19	[3]
23	[2, 1]
29	[2, 1]
37	[2, 1]
41	[3]
43	[2, 1]
47	[1, 1, 1]

2.2.2 A modular case

We can test reciprocity for f by comparing τ_p to p mod N for various N. Since the conductor and discriminant have the same prime factors N must be a power of 7. Let us first just take $N = 7$. We modify our function so that it also prints the values of p mod N,

```
pollisttype1(f,N,bd) =
{
  local(v);
  if(bd,,bd=50);

  forprime(p=2,bd,
      v=polfacttype(f,p);
      if(v,print(p,"\t",p%N,"\t",v)))
}
```

and we obtain the following output

```
? pollisttype1(x^3 + x^2 - 2*x - 1,7,100)
```

2	2	[3]
3	3	[3]
5	5	[3]
11	4	[3]
13	6	[1, 1, 1]
17	3	[3]
19	5	[3]
23	2	[3]
29	1	[1, 1, 1]
31	3	[3]
37	2	[3]
41	6	[1, 1, 1]
43	1	[1, 1, 1]
47	5	[3]
53	4	[3]
59	3	[3]
61	5	[3]
67	4	[3]
71	1	[1, 1, 1]
73	3	[3]
79	2	[3]
83	6	[1, 1, 1]
89	5	[3]
97	6	[1, 1, 1]

Staring at this table for a few minutes should convince us that τ_p is indeed modular of conductor 7; more precisely,

$$\tau_p = \begin{cases} [1,1,1] & \text{if } p \equiv \pm 1 \bmod 7 \\ [3] & \text{otherwise.} \end{cases} \qquad (2.9)$$

This is in fact true and can be proved by realizing that f is the minimal polynomial of $\zeta_7 + \zeta_7^{-1}$ where $\zeta_7 = \exp(2\pi i/7)$ is a primitive 7-th root of unity. (We confess that we obtained f with

```
algdep(trace(exp(2*Pi*I/7)),3)
```

see §1.2.3, though this

```
? minpoly(Mod(x+1/x, polcyclo(7)))
```

would have been the right thing to do.)

We can use the GP built-in function `nfisincl`, which checks whether the field defined by the first argument can be embedded into the field defined by the second; its output is either 0 if there is no embedding or a vector giving all the maps of one field to the other. Concretely, we find

```
? nfisincl(x^3+x^2-2*x-1,polcyclo(7))

[x^4 + x^3, -x^5 - x^4 - x^3 - x^2 - 1, x^5 + x^2]
```

Here `polcyclo(n)` returns the n-th cyclotomic polynomial (in the variable x unless specified otherwise).

2.2.3 A non-modular case

It is harder to convince ourselves numerically that τ_p is not modular for g. Recall, however, that by Theorem 1.1 part of $\tau_p = [n_1, \ldots, n_k]$ is always modular, namely, the quantity $\sum_j (j + n_j)$ mod 2. For g this means that the value of the Kronecker symbol $(\frac{-31}{p})$ distinguishes between the factorization types $[3], [1, 1, 1]$ and $[1, 2]$, but Theorem 2.3 implies that we may not separate the two types $[3]$ and $[1, 1, 1]$ by congruences on p alone. In particular, we may not answer the question of whether $g \equiv 0$ mod p has a root modulo p or not by knowing the value of p mod N for a fixed N. Naturally, we may ask, *can* we distinguish the two types $[3]$ and $[1, 1, 1]$ at all? We will answer this question in Chapter 3.

In any case, here is something we could try. Given N we search for two primes p and q such that $p \equiv q$ mod N and $\tau_p \neq \tau_q$. For example, with

```
getpq(N, bd = 200) =
{
  local(tp);

    forprime(p=2, bd,
      if(kronecker(-31,p) != 1, next);
      tp = polfacttype(x^3+x+1, p);
      forstep(q = p+N, bd, N,
        if(!isprime(q) || kronecker(-31,q) != 1, next);
        if(tp != polfacttype(x^3+x+1, q), return([p,q]))))
}
```

we find

```
? for(N=2,31,print(N,"\t",getpq(N)))

2       [5, 47]
3       [2, 47]
4       [5, 149]
5       [2, 47]
```

6	[5, 47]
7	[2, 149]
8	[5, 149]
9	[2, 47]
10	[7, 47]
11	[19, 173]
12	[5, 149]
13	[2, 67]
14	[5, 47]
15	[2, 47]
16	[5, 149]
17	[67, 101]
18	[5, 131]
19	[2, 173]
20	[7, 47]
21	[2, 149]
22	[19, 173]
23	[67, 113]
24	[5, 149]
25	[47, 97]
26	[19, 149]
27	[41, 149]
28	[5, 173]
29	[47, 163]
30	[7, 67]
31	[5, 67]

Of course, this only proves that if $Q[x]/(g)$ is modular its conductor must be at least 32. We can try only powers of 31

```
? getpq(31^2,10^4)

 [7, 3851]

? getpq(31^3,10^5)

 [47, 59629]
```

but for all we know its conductor is, say, $31^5 = 28629151$, which would be hard to disprove with our simpleminded approach. We should point out, however, that by a theorem of Tchebotarev discussed in the next section the function getpq is guaranteed to always find a pair of primes p and q for a large enough value of bd (since A is not in fact modular by Theorem 2.3). How large a bound bd is enough can be obtained from explicit versions of Tchebotarev's theorem, see [56].

2.3 The Artin map

To go into some more detail on the general reciprocity law we need to discuss the Artin map. Let $K = \mathbf{Q}[x]/(f)$ be a finite extension, let $L/\mathbf{Q}$ be its Galois closure and set $G := \operatorname{Gal}(L/\mathbf{Q})$. We have an action of G on the roots of the polynomial f, which after labeling them in some way, gives an embedding $\iota : G \subseteq S_n$, where $n = \deg(f) = [K : \mathbf{Q}]$.

Here S_n is the symmetric group of permutations of n objects. Recall that any $\sigma \in S_n$ has an essentially unique decomposition into disjoint cycles $\sigma = \eta_1 \cdots \eta_k$. If we reorder these cycles so that $n_1 \geq n_2 \geq \cdots \geq n_k$, where n_j is the length of η_j then $(n_1, n_2, \ldots, n_k)$ is a partition of n; i.e., $n_1 + \cdots + n_k = n$. The conjugacy class of σ in S_n is uniquely determined by this partition.

It is not hard to see (Ex. 2) that if $K = L$, i.e., if $K/\mathbf{Q}$ is Galois, then for any $g \in G$ the conjugacy class of $\iota(g)$ corresponds to the partition

$$(d, \ldots, d) \tag{2.10}$$

where d is the order of g.

The Artin map assigns to any unramified prime p a canonical conjugacy class Frob_p in G; under ι it maps to the conjugacy class of S_n determined by τ_p. In particular, the information contained in Frob_p is finer than that of τ_p since conjugacy classes in G may get identified when viewed in S_n via ι. (We will see this already in the small example at the end of §2.3.1 with $G \simeq D_4$.)

Rather than attempting to give more details on these claims (as the discussion would become too technical) let us consider how we can describe it in practice in a (small) example.

2.3.1 A Galois example

Let $f = x^8 + 2x^7 + 2x^6 - 2x^5 - 2x^4 - 2x^3 + 2x^2 + 2x + 1$. In this case $K = L$ and $G \simeq D_4$, the dihedral group of order 8 of symmetries of a square. (This example is considered in some more detail in §2.5 below.)

Assume $p \notin S_f \supseteq S_K$. If $f = f_1 \cdots f_r$ with $f_j \in \mathbf{F}_p[x]$ irreducible of degree n_j for $j = 1, \ldots, r$ then

$$\mathbf{F}_p[x]/(f) \simeq \mathbf{F}_{p^{n_1}} \times \cdots \times \mathbf{F}_{p^{n_r}},$$

where $\mathbf{F}_{p^{n_j}} \simeq \mathbf{F}_p[x]/(f_j)$.

Each of the finite fields extensions $\mathbf{F}_{p^{n_j}}/\mathbf{F}_p$ has an automorphism $x \mapsto x^p$ (called the *Frobenius automorphism*) which generates its Galois group. This automorphism arises from a unique $\sigma_j \in G$ of order n_j. We will spell out precisely what this means for this example below (2.12).

The σ_j are conjugate elements in G and, in fact, the set $\{\sigma_j\}$ is a full conjugacy class Frob_p in G (some of the σ_j for $j = 1, \ldots, r$ might coincide however). In particular, the σ_j have the same order d, say, and hence $d = n_1 = n_2 = \cdots = n_r$ (which is is consistent with (2.10)).

We compute the following

```
? f=x^8 + 2*x^7 + 2*x^6 - 2*x^5 - 2*x^4 - 2*x^3 + 2*x^2 +
2*x + 1;

? galv=nfgaloisconj(f)

[x, 1/2*x^6 + 1/2*x^5 - 2*x^3 - 1/2*x^2 - 1/2*x + 1, -x^7 -
2*x^6 - 2*x^5 + 2*x^4 + 2*x^3 + 2*x^2 - 2*x - 2, -1/2*x^7 -
x^6 - 1/2*x^5 + 2*x^4 + 3/2*x^3 - 3/2*x - 1, -1/2*x^7 -
1/2*x^6 + 2*x^4 + 1/2*x^3 + 1/2*x^2 - 2*x - 1, 1/2*x^7 -
1/2*x^5 - 2*x^4 + 3/2*x^3 - x^2 + 3/2*x - 1, 1/2*x^7 + x^6
+ x^5 - 3/2*x^4 - 3/2*x^3 - x^2 + 2*x + 1/2, x^7 + 2*x^6 +
3/2*x^5 - 5/2*x^4 - 2*x^3 + 3/2*x + 3/2]~
```

The function nfgaloisconj returns a vector with all the conjugate roots; i.e.,

$$f(g_k(x)) \equiv 0 \bmod f \qquad (2.11)$$

for each entry g_k of galv. We hence have an explicit way to describe G. (We should stress that doing this becomes unfeasible very quickly as $\deg(f)$ increases.)

In general, assuming without loss of generality that $f \in \mathbb{Z}[x]$ is monic, the denominator of any coefficient of a polynomial in nfgaloisconj divides m where $\mathrm{disc}(f) = m^2 \mathrm{disc}(K)$. In our example $m = 2^4$ (poldisc(f)/nfdisc(f) returns 256).

We claim that for each $p \notin S_f$ (this automatically excludes primes diving a denominator in galv by our previous discussion) and $j = 1, 2, \ldots, r$ there exists a unique $k = 1, 2, \ldots, 8$ such that

$$x^p \equiv g_k(x) \bmod f_j, \quad \text{in } \mathbf{F}_p[x]. \qquad (2.12)$$

This element g_k of galv represents $\sigma_j \in G$.

The following GP script will return an ordered list of the indices k such that g_k arises as above for some j

```
polallfrob(p,galv,f)=
{
 local(w,fv);

    if(!(poldisc(f)%p),return([]));
    fv=factormod(f,p)[,1];
    w = vector(length(fv), k, memb(Mod(x,fv[k])^p, galv % fv[k]));
    vecsort(w)
}
```

Note that

```
? factor(poldisc(f));

[2 20]

[5  4]
```

hence $S_f = \{2, 5\}$. We find

```
? forprime(p=7,100,print(p, "\t", polallfrob(p,galv,f)))
7       [2, 7]
11      [3, 3, 4, 4]
13      [5, 5, 8, 8]
17      [5, 5, 8, 8]
19      [3, 3, 4, 4]
23      [2, 7]
29      [1, 1, 1, 1, 1, 1, 1, 1]
31      [3, 3, 4, 4]
37      [5, 5, 8, 8]
41      [6, 6, 6, 6]
43      [2, 7]
47      [2, 7]
53      [5, 5, 8, 8]
59      [3, 3, 4, 4]
61      [6, 6, 6, 6]
67      [2, 7]
71      [3, 3, 4, 4]
73      [5, 5, 8, 8]
79      [3, 3, 4, 4]
83      [2, 7]
89      [1, 1, 1, 1, 1, 1, 1, 1]
97      [5, 5, 8, 8]
```

The set of elements $\{g_k\}$ in the output of allfrob for a prime p is a full conjugacy class in $G \simeq D_4$, namely Frob$_p$. To identify this class we need to know how the elements of D_4 are identified with the entries of galv (whose ordering, determined by the function nfgaloisconj, is basically random).

As a group of symmetries of a square in the plane, D_4 consists of the rotations a^k for $k = 0, \ldots, 3$ where a is, say, a counterclockwise rotation of angle $\pi/2$, and reflections $\sigma_j = a^j \sigma_0$ for $j = 0, \ldots, 3$ with respect to the four axis of symmetry.

It is not hard to work out that the conjugacy classes consist of

$$\{1\}, \{a^2\}, \{a, a^{-1}\}, \{\sigma_0, \sigma_2\}, \{\sigma_1, \sigma_3\}. \tag{2.13}$$

With a little bit more work we see that these conjugacy classes may be labeled with the indexing of galv as [1], [6], [2,7], [3,4], [5,8] respectively. Note however that the labeling of the last two classes requires that we make a choice since there is no intrinsic way to tell them apart.

Note also that we may recover τ_p from the output of allfrob but not viceversa: the outputs [6,6,6,6], [3,3,4,4], [5,5,8,8] all correspond to $\tau = [2,2,2,2]$. For comparison, here is the full output of types.

```
? pollisttype(f,100)
```

3	[4, 4]
7	[4, 4]
11	[2, 2, 2, 2]
13	[2, 2, 2, 2]
17	[2, 2, 2, 2]
19	[2, 2, 2, 2]
23	[4, 4]
29	[1, 1, 1, 1, 1, 1, 1, 1]
31	[2, 2, 2, 2]
37	[2, 2, 2, 2]
41	[2, 2, 2, 2]
43	[4, 4]
47	[4, 4]
53	[2, 2, 2, 2]
59	[2, 2, 2, 2]
61	[2, 2, 2, 2]
67	[4, 4]
71	[2, 2, 2, 2]
73	[2, 2, 2, 2]
79	[2, 2, 2, 2]
83	[4, 4]
89	[1, 1, 1, 1, 1, 1, 1, 1]
97	[2, 2, 2, 2]

Though the conjugacy class Frob_p gives finer information than the type τ_p, computationally the catch is that, in principle, we also need a description of the full Galois group (which could be as large as S_n for polynomials of degree n).

2.3.2 A non-Galois example

Using nfsubfields(f,4) we may find the five subfields of $L/\mathbf{Q}$ of the previous section §2.3.1. Let $K/\mathbf{Q}$ be the subfield given by $\mathbf{Q}[x]/(g) \hookrightarrow L$ with $g = x^4 + 2x^3 + 4x + 4$ and embedding defined by $x \mapsto -\frac{1}{2}x^7 - x^6 - \frac{1}{2}x^5 + 2x^4 + \frac{3}{2}x^3 - \frac{1}{2}x - 1$.

By our previous calculations we already know what Frob$_p$ is for $p = 7, 11, \ldots, 97$; we may now compare that with τ_p for g. We find the following

```
? g= x^4 + 2*x^3 +4*x + 4;
? pollisttype(g,100)
```

3	[4]
7	[4]
11	[2, 1, 1]
13	[2, 2]
17	[2, 2]
19	[2, 1, 1]
23	[4]
29	[1, 1, 1, 1]
31	[2, 1, 1]
37	[2, 2]
41	[2, 2]
43	[4]
47	[4]
53	[2, 2]
59	[2, 1, 1]
61	[2, 2]
67	[4]
71	[2, 1, 1]
73	[2, 2]
79	[2, 1, 1]
83	[4]
89	[1, 1, 1, 1]
97	[2, 2]

which is perfectly consistent: via the embedding $\iota : G \hookrightarrow S_4$ determined by K, galv[2] and galv[7] map to a 4-cycle; galv[6] to a product of two transpositions and galv[3], galv[4] fix two roots and galv[5], galv[8] fix none. (If we took the other non-Galois subfield of degree four of $L/\mathbf{Q}$ not conjugate to $K/\mathbf{Q}$ the last two cases would be interchanged, Ex. 3.)

Note that again knowing τ_p does not allow us to always recover Frob$_p$; for example, the classes $\{a^2\}$ and $\{\sigma_1, \sigma_3\}$ both map to the type $\tau = (2, 2)$.

Finally, we can actually compute the image of G in S_4 explicitly as follows. We first compute all the embeddings of $\mathbf{Q}[x]/(g) \hookrightarrow S_4$ with

```
? inclv=nfisincl(g,f);
```

We may think of the elements of inclv as the roots of g in L. If $g \in \mathbf{Q}[x]$ represent an element of G (an entry in galv) and $h \in \mathbf{Q}[x]$ represents such an embedding (an entry in inclv) then the action of g on h is represented by $u \in \mathbf{Q}[x]$ such that

$$u(x) \equiv h(g(x)) \bmod f.$$

Hence the following routine will list the elements $g \in G$ (ordered as in galv) viewed as permutations $\sigma \in S_4$ described as the vector $(\sigma(1), \sigma(2), \sigma(3), \sigma(4))$.

```
polgalperm(inclv,galv,f)=
{
  vector(length(galv),j,
    vector(length(inclv),k,
      memb(subst(inclv[k],x,Mod(galv[j],f)), inclv)))
}
```

We find

```
? polgalperm(inclv,galv,f)

[[1, 2, 3, 4], [4, 3, 1, 2], [2, 1, 3, 4], [1, 2, 4, 3],
 [4, 3, 2, 1], [2, 1, 4, 3], [3, 4, 2, 1], [3, 4, 1, 2]]
```

or in their cycle decomposition

1	1
2	(1423)
3	(12)
4	(34)
5	(14)(23)
6	(12)(34)
7	(1324)
8	(13)(24)

This matches our previous calculations. For example, this list shows that the elements of G labeled $5, 6, 8$ correspond to the partition $(2, 2)$ of 4. Hence any prime p whose Frobenius class is [6] or [5, 8] has factorization type $\tau_p = [2, 2]$ (see for example $p = 37$ and $p = 41$ in the output of polallfrob above).

2.4 Quantitative version

2.4.1 Application of a theorem of Tchebotarev

Let the notation be as in the previous section. We could ask what proportion of primes have a given factorization type. To make this more precise, given a partition $\tau = (n_1, \ldots, n_k)$ of n, we define

$$\pi(\tau, x) := \#\{p \leq x \mid \tau_p = \tau\} \tag{2.14}$$

and, with the usual notation,

$$\pi(x) := \#\{p \leq x\}. \tag{2.15}$$

A theorem of Tchebotarev implies that

$$\lim_{x \to \infty} \frac{\pi(\tau, x)}{\pi(x)} = \delta_\tau$$

where

$$\delta_\tau := \frac{\#C_\tau}{|G|}$$

with C_τ the union of the conjugacy classes in G which map via ι to the conjugacy class of S_n determined by τ.

In the case of §2.3.2 for example we know from our previous discussion that the possible C_τ's are:

$$C_{[1,1,1,1]} = \{1\}, \quad C_{[2,2]} = \{a^2, \sigma_0, \sigma_2\}, \quad C_{[4]} = \{a, a^{-1}\}, \quad C_{[2,1,1]} = \{\sigma_1, \sigma_3\}.$$

Therefore, the corresponding densities according to Tchebotarev are respectively:

$$\delta_{[1,1,1,1]} = 1/8, \quad \delta_{[2,2]} = 3/8 \quad \delta_{[4]} = 1/4, \quad \delta_{[2,1,1]} = 1/4. \tag{2.16}$$

2.4.2 Computing densities numerically

Let us test this numerically for the quartic field of §2.3.2. We may use the following GP function.

```
poltypedensity(f,t,e) =
{
    local(a,b,bd);

    if(e > 2,, e = 3); bd=10^e; e=bd/100;

    forprime(p=2,bd,
        if(polfacttype(f,p)==t,a=a++);
    b++;
    if(b%e,,print(a/b*1.)));

    print(a/b*1.)
}
```

The input for this function is a polynomial f defining the field, a type t, and an exponent e. The exponent is used to limit the number of primes used to

2.4 Quantitative version

10^e. In order to get an idea how the sequence $\pi(\tau,x)/\pi(x)$ is behaving but not wanting to see all of it, we print it only every so often (precisely, when the number b of primes which have been consider so far is a multiple of 10^{e-2} and at the very end). (WARNING: an error will occur if 10^e is bigger than the largest precomputed prime!)

Here are the results for the polynomial $g = x^4 + 2x^3 + 4x + 4$ according to type. (We first set the precision to nine digits as the output in the default precision of 28 digits only distracts.)

```
? g= x^4 + 2*x^3 + 4*x + 4;
? \p 9
   realprecision = 9 significant digits

? poltypedensity(g,[1,1,1,1],5)

0.116000000
0.118500000
0.122333333
0.121250000
0.121200000
0.122500000
0.123285714
0.123500000
0.122777778
0.122706422

? poltypedensity(g,[2,2],5)

0.378000000
0.374000000
0.376000000
0.375500000
0.375800000
0.375833333
0.374142857
0.374750000
0.375888889
0.375834028

? poltypedensity(g,[4],5)

0.253000000
0.256500000
0.250666667
0.252750000
0.250400000
```

```
0.249500000
0.249428571
0.250250000
0.249333333
0.250417014

? poltypedensity(g,[1,1,2],5)

0.251000000
0.250000000
0.250333333
0.250000000
0.252200000
0.251833333
0.252857143
0.251250000
0.251777778
0.250834028
```

The answers nicely match the limiting values given in (2.16).

Actually, rather than printing the rational number $\pi(\tau,x)/\pi(x)$ and trying to guess what its limiting value might be we can print instead its best rational approximation. We could also use `recognize` if we do not have a good idea of what N might be (see §1.2.3).

We can modify our function accordingly

```
poltypedensity1(f,t,N,e) =
{
  local(a,b,bd);
    if(e > 2,,e = 3); bd=10^e; e=bd/100;

    forprime(p=2,bd,
        if(polfacttype(f,p)==t,a=a++);
      b++;
        if(b%e,,print1(bestappr(a/b,N)," ")));
}
```

We now use the print command `print1` which does not go to the next line after printing so that we get all the numbers in a row instead of a column to save space. We now find

```
? poltypedensity1(g,[1,1,1,1],8,5)

1/8   1/8   1/8   1/8   1/8   1/8   1/8   1/8   1/8

? poltypedensity1(g,[2,2],8,5)
```

3/8 3/8 3/8 3/8 3/8 3/8 3/8 3/8 3/8

? poltypedensity1(g,[4],8,5)

1/4 1/4 1/4 1/4 1/4 1/4 1/4 1/4 1/4

? poltypedensity1(g,[2,1,1],8,5)

1/4 1/4 1/4 1/4 1/4 1/4 1/4 1/4 1/4

Similarly, for f as in §2.3.1 we find

? f=x^8 + 2*x^7 + 2*x^6 - 2*x^5 - 2*x^4 - 2*x^3 + 2*x^2 + 2*x + 1;

? poltypedensity1(f,[1,1,1,1,1,1,1,1],8,5)

1/8 1/8 1/8 1/8 1/8 1/8 1/8 1/8 1/8

? poltypedensity1(f,[2,2,2,2],8,5)

5/8 5/8 5/8 5/8 5/8 5/8 5/8 5/8 5/8

? poltypedensity1(f,[4,4],8,5)

1/4 1/4 1/4 1/4 1/4 1/4 1/4 1/4 1/4

This matches the fact that

$$C_{[1,1,1,1,1,1,1,1]} = \{1\}, \quad C_{[2,2,2,2]} = \{a^2, \sigma_0, \sigma_1, \sigma_2, \sigma_3\}, \quad C_{[4,4]} = \{a, a^{-1}\},$$

which we leave the reader to verify.

Experiments like these, however, quickly become unfeasible if we increase the degree. For example, if the Galois group is S_n then typedensity for $\tau = [1, 1, \ldots, 1]$ will approach $1/n!$ but it might take a long while before we even get a positive value! (See Ex. 4 and 5.)

2.5 Galois groups

Computing the Galois group of an irreducible polynomial $f \in \mathbf{Q}[x]$, i.e., the isomorphism class of the Galois group of the Galois closure of $\mathbf{Q}[x]/(f)$, is a hard problem. Already classifying all the groups that are possible for a given degree can be a daunting prospect and it has only been done completely for degrees ≤ 30. PARI-GP has a built in function polgalois that works for polynomials of degree at most 7 (and the galdata package this

2: Reciprocity

extends it up to degree 11). For example, for Trink's polynomial (see §2.5.2) $x^7 - 7x + 3$ we find

```
? polgalois(x^7-7*x+3)

[168, 1, 1, "L(7) = L(3,2)"]
```

The notation for the output of polgalois is by now a fairly standard one (see the PARI-GP manual); we find that $x^7 - 7x + 3$ has Galois group isomorphic to $PSL_2(F_7)$, the famous finite simple group of Klein of order 168.

If $Q[x]/(f)$ is Galois, f is Galois for short, then as we mentioned in §2.10 the factorization types $\tau_p = [n_1, \ldots, n_k]$ must have $n_1 = \cdots = n_k$. Using this we can write a simple test of whether f is Galois.

```
polisgalois(f, bd = 100 * poldegree(f))=
{
  local(v,k);

    forprime(p=2,bd,
      v = polfacttype(f,p);
      for (k=2, length(v), if (v[k] != v[k-1], return(0))));
    1
}
```

If the output of polisgalois is 0 we can be sure that the polynomial is not Galois; if the output is 1, however, we can only conclude that is is *likely* to be Galois. In this last case, we should run it again increasing the value of bd to have more confidence in the answer. (In theory there is a value of bd that guarantees that the answer is correct, it depends on estimates in Tchebotarev's density theorem, see [56], but we have not attempted to make this bound precise.) We can always try the more costly test:

```
length(nfgaloisconj(f)) == poldegree(f)
```

As an example we search for Galois polynomials of degree 8 of small height; to cut the search space we only consider irreducible monic polynomials with constant term equal to ±1. We use the following GP function

```
polgal8(bd) =
{
  local(v,f);

    forvec(u = concat([[[1,1], [0,bd]],
      vector(6,k,[-bd,bd]), [[0,0]]]),
      f=Pol(u);
      gal8test(f+1);
      gal8test(f-1))
}
```

2.5 Galois groups

where

```
gal8test(f)=
{
  if(polisirreducible(f) && polisgalois(f),
    print(f," ",factor(nfdisc(f))))
}
```

and obtain

```
? polgal8(2)

x^8 - x^6 + x^4 - x^2 + 1    [2, 8; 5, 6]
x^8 - x^4 + 1    [2, 16; 3, 4]
x^8 + 1    Mat([2, 24])
x^8 + x^7 - x^5 - x^4 - x^3 + x + 1    [3, 4; 5, 6]
x^8 + 2*x^7 + 2*x^6 - 2*x^5 - 2*x^4 - 2*x^3 + 2*x^2
    + 2*x + 1    [2, 12; 5, 4]
x^8 - 2*x^7 + 2*x^6 - 2*x^5 - 2*x^4 + 2*x^3 + 2*x^2
    + 2*x + 1    [2, 12; 3, 6]
```

Let us emphasize that the above list consists of the only polynomials within the search space, i.e., irreducible polynomials $f \in Q[x]$ of the form $f = x^8 + \sum_{k=1}^{7} u_k x^k \pm 1$ with $u_k \in Z$ and $|u_k| \leq 2$ for $k = 2, \ldots, 7$ and $0 \leq u_1 \leq 2$, which are *likely* to be Galois (we can see that they all define non-isomorphic fields as their discriminants are different). To check that they *are* Galois (not using the extension package of polgalois) we can proceed as follows. Take, for example, the fifth polynomial in the list and compute as in §2.3

```
? f=x^8 + 2*x^7 + 2*x^6 - 2*x^5 - 2*x^4 - 2*x^3 + 2*x^2 +
2*x + 1;

? galv=nfgaloisconj(f)
? length(galv)
    8
```

since there are $8 = \deg(f)$ automorphisms of the field $Q[x]/(f)$ it is Galois.

Just for fun let us actually verify that galv indeed consists of an explicit version of the Galois group of f; i.e., let us check that (2.11) is valid for the entries of galv.

```
?   for(k=1,8,print(k," ",subst(f,x,galv[k])%f))

1 0
2 0
3 0
4 0
5 0
```

```
6  0
7  0
8  0
```

We should point out that we could have done our search of Galois degree 8 polynomials directly testing the length of nfgaloisconj(f) (or the extension of polgalois for that matter) but, apart from the fact that the calculation is meant mainly as an illustration, this would be significantly slower than using polisgalois as we did.

The vector galv contains an explicit description of the Galois group G. What group of order 8 is it? To answer this question, we may check that it is not abelian, for example, using the function

```
galisabelian(galv,f) =
{
  local(j,k,c);

    for(j=1,length(galv)-1,
      for(k=j+1,length(galv),
        c=subst(galv[j],x,galv[k])-subst(galv[k],x,galv[j]);
        c=c%f;
          if(c,return(0))));
    1
}
```

and we may also compute the order of each element in G using

```
galorder(g,f) =
{
  local(k,gg);
  k=1;gg=g;
    while(k<1000 && gg!=x,
      gg=subst(gg,x,g)%f;k++);
  k
}
```

We find that

```
? for(k=1,8,print(k," ",galorder(galv[k],f)))

1  1
2  4
3  2
4  2
5  2
6  2
7  4
8  2
```

since there are only two elements of order 4 we must have that $G \cong D_4$, the dihedral group of order 8.

Remark Both functions `galisabelian` and `galorder` will be exceedingly slow for a more complicated `galv`. The problem are the substitutions like

```
subst(galv[k],x,galv[j])
```

The right way to tackle these is to reduce modulo a prime p not dividing disc(f) and work modulo a factor f_p of f mod (p, f_p). We leave this to the reader (Ex. 8). Also, the function `galisabelian` can be replaced by the more efficient

```
galisabelian1(f)=galoisisabelian(galoisinit(f))
```

2.5.1 Tchebotarev's theorem

The full strength of Tchebotarev's theorem is as follows. Instead of counting how many primes $p \leq x$ give a certain factorization type we count how many have Frob$_p$ in a fixed conjugacy class C of the Galois group G. Then, with the obvious change in notation,

$$\lim_{x \to \infty} \frac{\pi(C, x)}{\pi(x)} = \frac{\#C}{|G|}.$$

To illustrate this numerically we go back to the example of §2.3.1. Using the following simple modification of the function `poltypedensity1`

```
poltypedensity2(f,galv,t,N,e)=
{
  local(a,b,bd);

  if(e>2,,e=3);bd=10^e;e=bd/100;

  forprime(p=2,bd,
      if(polallfrob(p,galv,f)==t,a=a++);
    b++;
      if(b%e,,print1(bestappr(a/b,N)," ")));
}
```

we obtain

```
? f=x^8 + 2*x^7 + 2*x^6 - 2*x^5 - 2*x^4 - 2*x^3 + 2*x^2 +
2*x + 1;

? galv=nfgaloisconj(f);
? poltypedensity2(f,galv,[1,1,1,1,1,1,1,1],8,5)
```

```
1/8   1/8   1/8   1/8   1/8   1/8   1/8   1/8   1/8

? poltypedensity2(f,galv,[6,6,6,6],8,5)

1/8   1/8   1/8   1/8   1/8   1/8   1/8   1/8   1/8

? poltypedensity2(f,galv,[2,7],8,5)

1/4   1/4   1/4   1/4   1/4   1/4   1/4   1/4   1/4

? poltypedensity2(f,galv,[3,3,4,4],8,5)

1/4   1/4   1/4   1/4   1/4   1/4   1/4   1/4   1/4

? poltypedensity2(f,galv,[5,5,8,8],8,5)

1/4   1/4   1/4   1/4   1/4   1/4   1/4   1/4   1/4
```

again matching the sizes of the respective conjugacy classes as listed in (2.13).

2.5.2 Trink's example

We will now look more closely at the polynomial $f = x^7 - 7x + 3$; as we mentioned in the previous section, Trink found that the Galois group G of the Galois closure $L/\mathbf{Q}$ of f is isomorphic to $\mathrm{PSL}_2(\mathbf{F}_7)$, a simple group of order 168. In fact, Trink discovered this by factoring f for small primes.

```
? poltypes(x^7-7*x+3)

 [[7], [4, 2, 1], [3, 3, 1], [2, 2, 1, 1, 1]]
```

Clearly this search is not sufficient as we know that the type $[1,1,1,1,1,1,1]$ must occur (as it corresponds to the identity element of G). We try with a larger search space

```
? poltypes(x^7-7*x+3,2,5000)

 [[7], [4, 2, 1], [3, 3, 1], [2, 2, 1, 1, 1], [1, 1, 1, 1,
 1, 1, 1]]
```

Let us check what the densities for each type look like numerically.

```
? poltypedensity1(x^7-7*x+3,[7],10,5)

2/7   2/7   2/7   2/7   2/7   2/7   2/7   2/7   2/7

? poltypedensity1(x^7-7*x+3,[4,2,1],10,5)
```

```
1/4   1/4   1/4   1/4   1/4   1/4   1/4   1/4   1/4

? poltypedensity1(x^7-7*x+3,[3,3,1],10,5)

1/3   1/3   1/3   1/3   1/3   1/3   1/3   1/3   1/3

? poltypedensity1(x^7-7*x+3,[2,2,1,1,1],10,5)

1/9   1/8   1/8   1/8   1/8   1/8   1/8   1/8   1/8
```

Since $2/7 + 1/4 + 1/3 + 1/8 + 1/168 = 1$ it looks like we have all possible types with their respective densities. (Note that the density for $[1, 1, 1, 1, 1, 1, 1]$ is $1/168$ since we know $|G| = 168$; computing it numerically directly would require a fairly large number of primes.) Indeed, these are the right densities and types (see Ex. 10).

Remark Since $PSL_2(F_7)$ is simple there are no non-trivial Galois sub-extensions of L/Q (the Galois closure of the field defined by f). In particular, there is no abelian sub-extensions and hence by Artin's reciprocity law Theorem 2.3 there is no easy way to separate the factorization types at all (by means of extended congruences as we can do in a dihedral case for example, see Chapter 3).

2.5.3 An example related to Trink's

Suppose we knew τ_p for all but finitely many primes p. Could we then identify the algebra (up to isomorphism)? Perhaps surprisingly the answer is no. This phenomenon is an analogue for étale algebras of the celebrated 1966 question of M. Kac *Can you hear the shape of a drum?* In other words, can we identify the shape of a drum if we know all of its modes of vibration? As it turns out you cannot. In technical terms what we want to construct are isospectral non-isomorphic manifolds. In 1985 Sunada [92] gave a construction of such manifolds using finite groups. The key ingredient is described below for the group $PSL_2(F_7)$. See also [41].

In fact, the situation had already been considered, in the number theory setting, by F. Gassman [39] in 1926: there exists two non-isomorphic number fields with identical zeta function. Indeed, we will now construct another degree 7 field which has the exact same zeta function as that defined by Trink's polynomial, i.e., all Euler factors F_p as in (2.7) for every prime p coincide, without being isomorphic to it. It follows from Tchebotarev's theorem that any two fields with the same zeta function must have the same Galois closure. Hence we will need to look for our field inside L/Q; unfortunately we do not have L/Q in any explicit form so we will have to do some work.

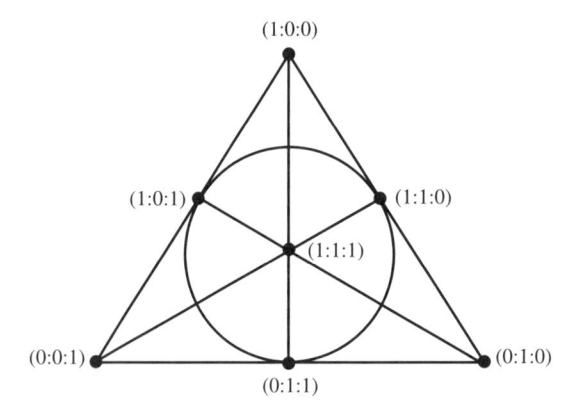

Figure 2.1 *Projective plane over F_2*

As we mentioned the Galois group G of the Galois closure of $f = x^7 - 7x + 3$ is isomorphic to $PSL_2(F_7)$. As it happens this group is (Ex. 9) also isomorphic to $GL_3(F_2) = PSL_3(F_2)$ and therefore acts on the projective plane $X := \mathbf{P}^2(F_2)$. This plane consists of seven points and seven lines with three points per line and three lines meeting at each point; it can be conveniently depicted as in Fig. 2.1.

The action of $PGL_3(F_2)$ on X mimics how the Galois group G acts on the roots of f. For example, the element of G represented by

$$\begin{pmatrix} 0 & 0 & 1 \\ 1 & 0 & 0 \\ 0 & 1 & 0 \end{pmatrix} \in GL_3(F_2)$$

which rotates the diagram clockwise a third of a turn, corresponds to the type $(3, 3, 1)$ (one fixed point and two 3-cycles).

Since the action of $PGL_3(F_2)$ on X preserves the incidence relations the seven sets of three points on each line will be permuted among each other. Suppose that we can identify the corresponding seven sets $S_1, S_2, \ldots, S_7$ of three roots of f and let η_j be the product of the three roots in S_j. Then

$$g := \prod_{j=1}^{7} (x - \eta_j) \tag{2.17}$$

will have rational coefficients. Unfortunately we do not really have an explicit identification of the roots of f with the points of X; so how can we compute g?

We could do the following. We first compute the polynomial $G_3(f)$ whose roots are all products of three distinct roots of f. In §7.2.3 we discuss a GP

2.5 Galois groups

routine `polwedge` which accomplishes this. We then look for g among the factors of $G_3(f)$.

```
? polwedge(x^7-7*x+3,3) [4]

x^35 - 7*x^33 - 1029*x^29 + 135*x^28 + 7203*x^27 - 756*x^26
+ 1323*x^24 + 352947*x^23 - 46305*x^22 - 2463339*x^21 +
324135*x^20 - 30618*x^19 - 453789*x^18 - 40246444*x^17 +
282225202*x^15 - 44274492*x^14 + 155098503*x^12 +
12252303*x^11 + 2893401*x^10 - 171532242*x^9 + 6751269*x^8
+ 2657205*x^7 - 94517766*x^6 - 3720087*x^5 + 26040609*x^3 +
14348907

? factor(%)

[x^7 + 7*x^6 + 21*x^5 + 28*x^4 - 14*x^3 - 63*x^2 + 27 1]

[x^28 - 7*x^27 + 21*x^26 - 28*x^25 - 35*x^24 + 210*x^23 -
1127*x^22 + 5498*x^21 - 15561*x^20 + 24437*x^19 - 14*x^18 -
69132*x^17 + 275779*x^16  - 1132390*x^15 + 2998617*x^14 -
5156389*x^13 + 3645859*x^12 - 130914*x^11 + 992439*x^10 +
2995461*x^9 + 845397*x^8 + 6008418*x^7 + 1183896*x^6 +
2985255*x^5 + 2342277*x^4 + 1240029*x^3 + 1240029*x^2 +
531441 1]
```

Since there is only one factor of degree 7 it must be g. The fields defined by f and g are not isomorphic but nevertheless they have the same zeta function.

We first check that the fields are not isomorphic.

```
? f=x^7-7*x+3;
? g=x^7 + 7*x^6 + 21*x^5 + 28*x^4 - 14*x^3 - 63*x^2 + 27;
? nfisisom(f,g)
0
```

Now we check that all factorization types τ_p for $2 \geq p \geq 100, p \neq 7$ agree.

```
? forprime(p=2,100,print(p, "\t", polfacttype(f,p), "\t",
polfacttype(g,p)))

2          [7]         [7]
3          []          []
5          [7]         [7]
7          []          []
11         [7]         [7]
13         [4, 2, 1]         [4, 2, 1]
17         [3, 3, 1]         [3, 3, 1]
19         [3, 3, 1]         [3, 3, 1]
23         [3, 3, 1]         [3, 3, 1]
```

29	[7]	[7]
31	[7]	[7]
37	[4, 2, 1]	[4, 2, 1]
41	[4, 2, 1]	[4, 2, 1]
43	[3, 3, 1]	[3, 3, 1]
47	[4, 2, 1]	[4, 2, 1]
53	[7]	[7]
59	[4, 2, 1]	[4, 2, 1]
61	[4, 2, 1]	[4, 2, 1]
67	[3, 3, 1]	[3, 3, 1]
71	[3, 3, 1]	[3, 3, 1]
73	[4, 2, 1]	[4, 2, 1]
79	[2, 2, 1, 1, 1]	[2, 2, 1, 1, 1]
83	[4, 2, 1]	[4, 2, 1]
89	[4, 2, 1]	[4, 2, 1]
97	[3, 3, 1]	[3, 3, 1]

This last calculation does not, of course, prove that the fields have the same zeta function though it makes it very likely. We sketch the main ingredients of a proof (see [92], [71] for more details). The fields defined by f and g can be embedded inside the same Galois closure $L/\mathbf{Q}$ with Galois group G (see [21] Exercise 6.1). By Galois theory they are the fixed fields of two subgroups of G, say U and V respectively. In the $\mathrm{PGL}_3(\mathbf{F}_2)$ version of G we can identify these subgroups as the stabilizers of a point and a line respectively. More precisely, U can be thought of as those matrices of the form

$$\begin{pmatrix} * & * & * \\ 0 & * & * \\ 0 & * & * \end{pmatrix} \in GL_3(\mathbf{F}_2)$$

and V of those of the form

$$\begin{pmatrix} * & 0 & 0 \\ * & * & * \\ * & * & * \end{pmatrix} \in GL_3(\mathbf{F}_2).$$

These subgroups are not conjugate (hence the fields are not isomorphic) but there is a bijection $\phi : U \longrightarrow V$, given by transposition $u \mapsto u^t$, which has the property that u and $\phi(u)$ are always conjugate of each other. (To be sure: for each u there is a $g \in G$ such that $gug^{-1} = u^t$; there just is no g that works for all u.) It follows that for each conjugacy class C of G we have

$$\#(C \cap U) = \#(C \cap V). \tag{2.18}$$

This is the key condition of Sunada's construction. It is equivalent to the equality of the zeta functions of the corresponding fixed fields (see exercise 6.4 in [21] and [71], Theorem 1). The first example of this phenomenon (non-isomorphic number fields with the same zeta function) is due to F. Gassman [39] where the fields have degree 180 ($G = S_6$, $U = \{1, (12)(34), (13)(24), (14)(23)\}$ and $V = \{1, (12)(34), (12)(56), (34)(56)\}$. Perlis proved [71] that the smallest degree possible is 7 (as in the example discussed here; see however Ex. 17).

2.6 Exercises

1. Find a polynomial $f \in \mathbf{Z}[x]$ and $m \in \mathbf{N}$ such that the number of solutions to the congruence $f(x) \equiv 0 \mod m$ is larger than $\deg(f)$.
2. Prove that if $K/\mathbf{Q}$ is Galois, then for any $g \in G$ the conjugacy class of $\iota(g)$ corresponds to the partition $(d, \ldots, d)$ where d is the order of g.
   More generally, if $\iota(g)$ corresponds to the partition $(n_1, n_2, \ldots, n_r)$ prove that $\operatorname{lcm}(n_1, n_2, \ldots, n_r)/n_i$ divides $[L : K]$ for each i, where $L/\mathbf{Q}$ is the Galois closure of $K/\mathbf{Q}$.
3. Repeat the calculations of §2.3.2 for all the degree four subfields of $L/\mathbf{Q}$ and check that the resulting ouputs are all consistent.
4. Consider the function $b(x) := \#\{p \leq x \mid p \equiv 3 \mod 4\} - \#\{p \leq x \mid p \equiv 1 \mod 4\}$. Tchebotarev's density theorem guarantees that $b(x)/\pi(x) \to 0$ as $x \to \infty$. Write a GP version of this function and study its graph.
5. Modify the routine `poltypes` so that it also records the first prime with the given type. Find those primes in the case of Trink's polynomial and other (interesting, favorite, random,...) polynomials.
6. Analyze the rest of the (likely to be) Galois degree 8 polynomials listed in the text. (Note that the first four are in fact cyclotomic polynomials.)
7. Verify that $x^8 - 8x^7 + 8x^6 + 40x^5 - 68x^4 - 16x^3 + 56x^2 - 16x + 1$ is Galois and find the structure of its Galois group.
8. Write an improved version of `galorder` implementing the idea mentioned in the remark at the end of §2.5.
9. Prove that $\operatorname{PGL}_3(\mathbf{F}_2) \simeq \operatorname{PSL}_2(\mathbf{F}_7)$.
10. Check that the types and densities found numerically for Trink's example are indeed correct.
11. Find a bijection of the roots of $x^7 - 7x + 3$ and the points on the projective plane $\mathbf{P}(\mathbf{F}_2)$, compatible with the actions of the Galois group and $\operatorname{PGL}_3(\mathbf{F}_2)$, respectively.
12. Show that the field defined by the polynomial $x^8 - 4x^7 + 7x^6 - 7x^5 + 7x^4 - 7x^3 + 7x^2 + 5x + 1$ has Galois closure isomorphic to that of Trink's polynomial.
13. Does knowing how many elements have a given order determine the structure of a finite group?
14. Suppose that a polynomial f has only the following factorization types $[7], [4, 2, 1], [3, 3, 1], [2, 2, 1, 1, 1], [1, 1, 1, 1, 1, 1, 1]$. Prove that the Galois group of f has trivial center.

15. Find the minimal polynomial $f \in \mathbf{Q}[x]$ of $\sqrt{2}+\sqrt{3}$. Show that f, though irreducible over $\mathbf{Q}$, is reducible modulo p for all primes p.
16. With f as in the previous exercise find an embedding $\mathbf{Q}[x]/(f) \subset \mathbf{Q}(\zeta_m)$ for some $m \in \mathbf{N}$.
17. Show that the following non-isomorphic degree 6 étale algebras have the same zeta function (see [72]).
$$\mathbf{Q} \times \mathbf{Q} \times \mathbf{Q}(\sqrt{2},\sqrt{3}), \qquad \mathbf{Q}(\sqrt{2}) \times \mathbf{Q}(\sqrt{3}) \times \mathbf{Q}(\sqrt{6}).$$

3 Positive definite binary quadratic forms

A significant portion of Gauss's *Disquitiones Arithmeticae* is about binary quadratic forms. Since then there have been numerous treatments, especially for the positive definite case. The most relevant reference for us will be Cox's book [26]. Also many of the routines described here have built-in versions in PARI. Nevertheless, because binary forms are so central to number theory, we include this short (and rather sketchy) chapter with the most basic facts and algorithms for the positive definite case. We include also a brief discussion of the Artin reciprocity law for imaginary quadratic fields, answering in the process the question raised in §2.2.3 of whether we may be able to separate the factorization types [3] and [1, 1, 1] for the polynomial $g = x^3 + x + 1$.

3.1 Basic facts

A *binary quadratic form* $Q = (a, b, c)$ is a form

$$Q(x, y) = ax^2 + bxy + cy^2, \quad a, b, c \in \mathbf{Z}.$$

Its *discriminant* is the quantity $D := b^2 - 4ac$. Note that $D \equiv 0, 1 \bmod 4$ and, conversely, any $D \equiv 0, 1 \bmod 4$ is the discriminant of some binary form Q, as is easily checked. The form Q is *positive definite* if and only if $Q(x, y) > 0$ for all non-zero $(x, y) \in \mathbf{Z}^2$; equivalently, $a > 0$ and $D < 0$. We say that Q is *primitive* if (a, b, c) have no common factor.

The group $SL_2(\mathbf{Z})$ of 2×2 integer matrices of determinant 1 acts on the set of binary quadratic forms by change of variables; the discriminant and the properties of being primitive and positive definite are preserved under this action. Two forms Q_1, Q_2 are *equivalent* if one is obtained from the other by such a change of variables; we will denote this by $Q_1 \sim Q_2$. The number of equivalence classes of primitive, positive definite, binary quadratic forms of discriminant D is finite; its number is called the *class number* and is denoted by $h(D)$.

3.1.1 Reduction

From now on we restrict ourselves to primitive, positive definite, binary quadratic forms. Such a form $Q = (a, b, c)$ is *reduced* if

$$a \le c, \quad |b| \le a \quad \text{and} \quad b \ge 0 \text{ in the case of equality in either condition.}$$

Every Q is equivalent to a unique reduced form Q_0 and there is an efficient algorithm to find it.

The reduction algorithm is very simple and beautiful. It works as follows. We start with $Q = (a, b, c)$, the form we want to reduce, and check whether $|b| \le 2a$ is satisfied; if it is not, we change Q to $(a, b-2na, c')$ for appropriate $n, c' \in \mathbf{Z}$ (corresponding to the action of the matrix $T^n = \begin{pmatrix} 1 & n \\ 0 & 1 \end{pmatrix}$) so that it does hold. Next we check whether $a < c$ holds; if it does not, we change Q to $(c, -b, a)$ (corresponding to action of the matrix $S = \begin{pmatrix} 0 & 1 \\ -1 & 0 \end{pmatrix}$) and start again; if it does, we are essentially done, we just need to make sure the boundary condition $b \le 0$ holds by applying T or T^{-1} if necessary.

We can visualize this algorithm as follows. To a form $Q = (a, b, c)$ of discriminant D we associate the point $z_Q := (-b + \sqrt{D})/2a$ in the upper half-plane. Reducing the form is equivalent to using the Möbius transformations $S : z \mapsto -1/z$ and $T : z \mapsto z + 1$ to move the point z_Q to the standard fundamental domain for the action of $SL_2(\mathbf{Z})$. See [83] for details and proofs.

Here is a version of this algorithm in GP (we leave it to the reader to verify that it performs correctly). A form $Q = (a, b, c)$ is given as a triple [a,b,c]. We input Q and obtain as output a vector with two components; the first is the reduced form Q_0 and the second is the bottom row (γ, δ) of the $SL_2(\mathbf{Z})$ matrix that achieves the reduction, which has the property that $Q_0(\gamma, \delta) = a$.

```
bqfred(q) =
{
  local(top,bot, a=q[1], b=q[2], c=q[3]);
  top=[-b,1]; bot=[shift(a,1),0];
    while(1,
      applyTpow(); if (c >= a, break);
      applyS());
    if(a == c && b > 0, applyS());

  bot[1] = (bot[1]-b*bot[2])/(2*a);
  [[a,b,c], bot]
}
```

where

```
bqfn(a,b)=ceil(1/2+b/shift(a,1))-1;

applyTpow() =
{
  local(aux, n = bqfn(a,b));
  if (!n, return);
  aux=a*n; c+=(aux-b)*n; b-=shift(aux,1);
  top+=n*bot;
}
```

and

```
applyS() =
{
  local(aux);
  aux=c; c=a; a=aux;
  b=-b;
  aux=top; top=-bot; bot=aux;
}
```

3.1.2 Cornachia's algorithm

As an application of the previous section we may now discuss the algorithm of Cornachia for finding a solution to $p = x^2 + y^2$, where $p \equiv 1 \mod 4$ is a prime number (we alluded to this in §1.4.4)

The idea is the following. Assume we can find a form $Q = (p, b, c)$ of discriminant -4; we then reduce it and find $\gamma, \delta \in \mathbf{Z}$ such that $Q_0(\gamma, \delta) = p$; since there is only one reduced form of discriminant -4, namely $x^2 + y^2$, we have solved the problem. How do we find Q? It amounts to finding a solution to the congruence $b^2 \equiv -4 \mod p$, which always has one (note, for example that $(\mathbf{Z}/p\mathbf{Z})^\times$ is cyclic of order divisible by 4). Here is how the whole thing would look like in GP (compare with fermat of §1.4.4).

```
fermat1(p) =
{
  local(b);
  b=lift(sqrt(Mod(-1,p)));
  b=bqfred([p,2*b,(b^2+1)/p])[2];
  vecsort(abs(b))
}
```

(the last line is purely cosmetic). Note that just as fermat we make no effort to verify that the input is appropriate; for comparison we have

```
? fermat1(10^50+577)

[64862689068739216422454245424, 76110653438083542245450401]
```

Here is a quick check that fermat1 works fine. Consider the GP script

```
checkfermat1(n) =
{
  local(p,v);
  until(p%4==1,p=nextprime(n++));
  v=fermat1(p);
  [p,v,p-(v[1]^2+v[2]^2)]
}
```

Given n it finds the first prime $p > n$ with $p \equiv 1 \bmod 4$ and outputs $(p, (x, y), p - (x^2 + y^2))$. Hence in our tests we better always have the last entry of the ouput equal 0. Here is an example.

```
? for(k=20,30,print(k," ",checkfermat1(10^k)))
20  [100000000000000000129, [4418521500, 8970878873], 0]
21  [1000000000000000000117, [21329289926, 23346549879], 0]
22  [10000000000000000000009, [3, 100000000000], 0]
23  [100000000000000000000117, [99718464234, 300093698519],
0]
24  [1000000000000000000000049, [7, 1000000000000], 0]
25  [10000000000000000000000013, [1007150657747,
2997606971002], 0]
26  [100000000000000000000000273, [551284805508,
9984792690047], 0]
27  [1000000000000000000000000301, [20697728985875,
23908241567026], 0]
28  [10000000000000000000000000457, [21612544638856,
97636560335939], 0]
29  [100000000000000000000000000481, [26262982594615,
315135297523516], 0]
30  [1000000000000000000000000000057, [407947858332109,
913005227193276], 0]
```

Remark Note that the algorithm can be modified to compute a solution to

$$n = x^2 + y^2$$

for any n starting from a solution to the congruence $b^2 \equiv -1 \bmod n$.

3.1.3 Class number

Nowadays there exist fast and sophisticated algorithms for computing the class number $h(D)$ (for D positive or negative). For $D < 0$ not too large one can use the theory of Gauss reduction of the previous section to compute $h(D)$ by simply enumerating all possible reduced forms of discriminant D; in many cases, this is just what we need. Here is a GP version of the enumeration process. (Thanks to J. Tate for suggesting a significant improvement in the code.)

```
bqf(D) =
{
  local(b0 = D%2, fv,a,c,zv);

  if(D >= 0 || D%4 > 1, return([]));

  fv = [[1,b0,(b0^2-D)/4]];
  forstep(b = b0, floorsqrt(-D/3), 2,
      zv=divisors((b^2-D)/4);
      n=length(zv);
      forstep(j=(n+1)\2, 2, -1,
        a=zv[j]; if (a < b, break);
        c=zv[n-j+1];
        if(gcd(gcd(a,b),c) != 1, next);
        fv = concat(fv,[[a,b,c]]);
        if(b && a != b && a != c, fv = concat(fv,[[a,-b,c]])))));
  fv
}
```

The output of bqf is simply a vector whose entries are all the reduced forms of the given discriminant (or the empty vector [] if the input is not a negative discriminant). For example,

```
? bqf(-31)

[[1, 1, 8], [2, 1, 4], [2, -1, 4]]

? bqf(-47)

[[1, 1, 12], [3, 1, 4], [3, -1, 4], [2, 1, 6], [2, -1, 6]]

? bqf(-100)

[[1, 0, 25], [2, 2, 13]]
```

To check the function, we can use the following GP script (which should have no output!)

```
checkbqf(x) =
{
  for(d=x,x+1000,
    v=bqf(-d);if(v,if(length(v)-qfbclassno(-d),print(-d))))
}
```

3.1.4 Composition

Gauss discovered that binary quadratic forms can be *composed* (see [26] Chapter I, §3 A). The operation of composition on forms gives a structure of a (finite) abelian group to the set of primitive positive definite quadratic forms of a given discriminant D. We will denote this group by $Cl(D)$ and call it the *class group* of forms of discriminant D.

Technical remark Given a discriminant $D < 0$ there is a unique order $\mathcal{O}$ in the imaginary quadratic field $K = \mathbb{Q}(\sqrt{D})$ of discriminant D. There is a natural bijection between binary quadratic forms and invertible ideals in $\mathcal{O}$. Via this association, $Cl(D)$ corresponds to the class group of $\mathcal{O}$. More precisely $Cl(D) = \text{Pic}(\mathcal{O})$, the group of invertible ideals modulo principal ones. For a fundamental discriminant D the order $\mathcal{O}$ is the full ring of integers of K. The corresponding group $Cl(D)$ is called the *class group* of K and, as is customary, we will also denote it $Cl(K)$.

Here is a GP routine for computing the composition of forms.

```
bqfcomp(q1,q2) =
{
  local(aux,m,a,b,c,dd,n);

  dd=-bqfdisc(q1);

  if(bqfdisc(q2) != -dd,return([]));

  aux=bezout(q1[1],q2[1]);
  m=aux[2];
  aux=bezout(aux[3],(q1[2]+q2[2])\2);
  m=m*aux[1];n=aux[2];

  a=q2[1]\aux[3];
  b=q2[2]+(m*(q1[2]-q2[2])\2-n*q2[3])*a*2;
  a=a*q1[1]\aux[3];
  b=b%(2*a);
  c=(b*b+dd)\(4*a);

  bqfred([a,b,c])[1]
}
```

where for convenience we define

```
bqfdisc(q)=q[2]^2-shift(q[1]*q[3],2)
```

Under composition the form $(1, *, *)$ represents the identity element and the inverse of the class of (a, b, c) is the class of $(a, -b, c)$. In particular, elements of order 2 are precisely those represented by reduced forms with shape (a, b, a), (a, a, c), or $(a, 0, c)$ and $a > 1$ (note that in all cases $a \mid D$). If $D = -p$ with $p \equiv 3 \bmod 4$ an odd prime then $CL(K)$ is odd (see Ex. 5).

The following function computes the successive composition of a form with itself (reducing at each stage). We assume the input satisfies $n \geq 0$.

```
bqfpow(q,n)=
{
  local(t,dd);
    if(n == 0, dd=-bqfdisc(q); return([1,dd%2,(dd+dd%2)/4]));
    if(n == 1, return (q));
  t = bqfpow(bqfcomp(q,q), n\2);
    if(n % 2, bqfcomp(q,t), t);
}
```

For example, $Cl(-47)$ is cyclic of order 5 hence any non-trivial form is a generator. Let us check this.

```
? vector(5,k,bqfpow([2,1,6],k))

[[2, 1, 6], [3, -1, 4], [3, 1, 4], [2, -1, 6], [1, 1, 12]]
```

3.2 Examples of reciprocity for imaginary quadratic fields

3.2.1 Dihedral group of order 6

We now briefly explain how Artin's reciprocity law for the imaginary quadratic field $K = \mathbf{Q}(\sqrt{-31})$ allows us to distinguish between the factorization types [3] and [1, 1, 1] for the polynomial $g = x^3 + x + 1$ of §2.2.3. Unfortunately, the discussion requires more knowledge of algebraic number theory than we have been assuming until now.

The Galois closure $H/\mathbf{Q}$ of $\mathbf{Q}[x]/(g)$ is an unramified cubic extension of K. Since $h(-31) = 3$

```
? bqf(-31)

  [[1, 1, 8], [2, 1, 4], [2, -1, 4]]
```

we deduce that H is the Hilbert class field of K; by Artin's reciprocity law for K the Artin map gives an isomorphism $\Phi : Cl(K) \longrightarrow \mathrm{Gal}(H/K)$, where $Cl(K)$ is the class group of K (see §3.2.5 for more on this). Suppose that p is a prime which splits in K as $p = \mathcal{P}\bar{\mathcal{P}}$, i.e., $(\frac{-31}{p}) = 1$. Then $\Phi(\mathcal{P}) \in \text{Gal}(H/K)$ is trivial if and only if the class of $\mathcal{P}$ is trivial in $Cl(K)$. The class of $\mathcal{P}$ in $Cl(K)$ is that of a binary quadratic form $Q_p = (p, *, *)$ of discriminant -31. Since $\Phi(\mathcal{P})\Phi(\bar{\mathcal{P}}) = 1$ to know if either one of the factors is trivial in $\text{Gal}(H/K)$ we just need to check whether $Q_p \sim (1, -1, 8)$ for *any* (there are two for $p \neq 31$) such form Q_p.

The missing ingredient that will now allow us to distinguish the types [3] and [1, 1, 1] is precisely that condition. For p such that $(\frac{-31}{p}) = 1$ we have

$$\tau_p = [1, 1, 1] \iff Q_p \sim (1, -1, 8). \tag{3.1}$$

We can reformulate this further since $Q_p \sim (1, -1, 8)$ if and only if $p = x^2 - xy + 8y^2$ for some $x, y \in \mathbb{Z}$. Indeed, just like in Cornachia's algorithm §3.1.2, we find such x and y in the process of reducing the form Q_p. Hence,

$$\tau_p = [1, 1, 1] \iff p = x^2 - xy + 8y^2, \quad x, y \in \mathbb{Z}. \tag{3.2}$$

This is a truly remarkable generalization of quadratic reciprocity: $x^3 + x + 1$ is irreducible modulo p, when $(\frac{-31}{p}) = +1$, if and only if p cannot be written as $p = x^2 - xy + 8y^2$ for some integers x and y. (Compare with [26] Theorem 0.5). We may say that such conditions on p are a sort of *extended congruence* conditions. A part of Artin's general reciprocity law roughly says that for an arbitrary number field F the splitting of primes in an abelian extension L/F is determined by analogous extended congruence conditions. Unfortunately, it is hard in general to make these conditions as explicit as we did for an imaginary quadratic field. It is a beautiful, but also potentially misleading, accident that these conditions in the case for an imaginary quadratic field turn out to involve the classical theory of binary quadratic forms.

Let us see how it works numerically.

```
? forprime(p=2,150,if(kronecker(-31,p)==1,print(p, "\t",
bqfred(bqfprimef(p,-31))[1], "\t", polfacttype(x^3+x+1,p))))

2         [2, 1, 4]      [3]
5         [2, -1, 4]     [3]
7         [2, 1, 4]      [3]
19        [2, -1, 4]     [3]
41        [2, -1, 4]     [3]
47        [1, 1, 8]      [1, 1, 1]
59        [2, -1, 4]     [3]
67        [1, 1, 8]      [1, 1, 1]
71        [2, -1, 4]     [3]
97        [2, -1, 4]     [3]
101       [2, 1, 4]      [3]
```

103	[2, 1, 4]	[3]
107	[2, 1, 4]	[3]
109	[2, -1, 4]	[3]
113	[2, 1, 4]	[3]
131	[1, 1, 8]	[1, 1, 1]
149	[1, 1, 8]	[1, 1, 1]

Here we used the function

```
bqfprimef(p,D) =
{
  local(b = centerlift(sqrt(Mod(D,p))));
    if((b-D)%2,b=b+p);
  [p,b,(b^2-D)/4/p]
}
```

to construct one of the forms $Q_p = (p, *, *)$ of discriminant D (compare with `fermat1`).

To cast a wider net we can try this

```
checkcubic(bd) =
{
  forprime(p=2, bd,
    if(kronecker(-31,p)!=1, next);
    if((polfacttype(x^3+x+1,p)==[1,1,1]) != bqfistrivial(p,-31),
       print(p)))
}
```

where

```
bqfistrivial(p,D) =
{
  local(q);
  q=bqfprimef(p,D);
    if(bqfred(q)[1]==1,1,0)
}
```

is used to check whether a form Q_p is equivalent to the trivial one or not. Running `checkcubic` with any input should produce no output.

3.2.2 Theta functions again

There is one final reformulation of the reciprocity law for the dihedral examples 3.2.1 and 3.2.3.

As we did in §1.4.5, to any positive definite binary quadratic form Q we associate its *theta function*

$$\Theta_Q = \sum_{x,y \in \mathbb{Z}} q^{Q(x,y)} = \sum_{n \geq 0} r_Q(n) q^n, \qquad (3.3)$$

whose coefficients $r_Q(n)$ count the number of solutions in integers to $n = Q(x, y)$. Clearly two equivalent forms have the same theta function as the number of representations of any integer is the same for both forms.

Consider

$$F := \tfrac{1}{2}\left(\Theta_{(1,-1,8)} - \Theta_{(2,1,4)}\right) = \sum_{n \geq 1} a_n q^n \qquad (3.4)$$

(note that $\Theta_{(2,1,4)} = \Theta_{(2,-1,4)}$ so we are not really ignoring the form $(2, -1, 4)$).

Numerically we can easily compute the first few coefficients of the expansion of F as follows

```
? hv=bqf(-31);
? F=(bqftheta(hv[1],150)-bqftheta(hv[2],150))/2
```

$q - q^2 - q^5 - q^7 + q^8 + q^9 + q^{10} + q^{14} - q^{16} - q^{18} - q^{19} + q^{31} + q^{35} + q^{38} - q^{40} - q^{41} - q^{45} + 2*q^{47} - q^{56} - q^{59} - q^{62} - q^{63} + q^{64} + 2*q^{67} - q^{70} - q^{71} + q^{72} + q^{80} + q^{81} + q^{82} + q^{90} - 2*q^{94} + q^{95} - q^{97} - q^{101} - q^{103} - q^{107} - q^{109} + q^{112} - q^{113} + q^{118} + q^{121} + q^{125} + q^{126} - q^{128} + 2*q^{131} + q^{133} - 2*q^{134} + q^{142} - q^{144} + 2*q^{149} + O(q^{150})$

where

```
bqftheta(Q,bd) =
{
  local(bd);
  b=sqrtint(bd);
  sum(n=-b,b,
    sum(m=-b,b,
      q^bqfeval(Q,[m,n]))+O(q^bd))
}
```

with

```
bqfeval(q,v)=q[1]*v[1]^2+q[2]*v[1]*v[2]+q[3]*v[2]^2
```

computes Θ_Q to the desired order of precision (we gave no thought to do this cleverly; we just want to make sure the answer is correct).

We can rephrase Artin's reciprocity in terms of F: the type τ_p and the coefficient a_p of F are correlated as follows

τ_p	a_p
[1, 1, 1]	2
[2, 1]	0
[3]	−1

3.2 Examples of reciprocity for imaginary quadratic fields

and indeed numerically we find

```
? f=x^3+x+1;
? forprime(p=2,100,print(p, "\t", polcoeff(F,p), "\t",
polfacttype(f,p)))

2       -1      []
3       -1      []
5       0       [2, 1]
7       -1      [3]
11      0       [2, 1]
13      0       [3]
17      -1      [2, 1]
19      0       [3]
23      0       [2, 1]
29      0       [2, 1]
31      0       [1, 1, 1]
37      -1      [3]
41      0       [2, 1]
43      0       [1, 1, 1]
47      2       [2, 1]
53      -1      [2, 1]
59      -1      [2, 1]
61      -1      [3]
67      0       [3]
71      -1      [2, 1]
73      0       [3]
79      -1      [3]
83      4       [2, 1]
89      -1      [2, 1]
97      -1      [3]
```

Technical remark Alert readers may have noticed in the above table that the correspondence between τ_p and a_p is more precisely given as follows: a_p is one less than the number of 1's in τ_p. Here is why. The generating function F in (3.4) is a newform of weight 1 of level 31 and character $\left(\frac{-31}{\cdot}\right)$. Its associated L-function $L(F,s)$ equals that of the irreducible Artin representation: $\rho : \mathrm{Gal}(\overline{\mathbf{Q}}/\mathbf{Q}) \longrightarrow \mathrm{Gal}(H/\mathbf{Q}) \longrightarrow \mathrm{GL}_2(\mathbf{C})$, where the last map is determined by the natural action of the dihedral group $\mathrm{Gal}(H/\mathbf{Q})$ on the plane. Let χ be the character of ρ, then $\chi(\mathrm{Frob}_p) = a_p$. On the other hand, we have the permutation representation ρ' of $\mathrm{Gal}(\overline{\mathbf{Q}}/\mathbf{Q})$ on the roots of g. Its character gives the number of fixed points of the action, i.e. the number of 1's in τ_p, and it is easy to see that $\rho' = \rho \oplus 1$, where 1 denotes the trivial representation. For this specific example see [84] §7.3.

What we have here is a relation between a two-dimensional Galois representation and a modular form of weight 1. In this case of a dihedral group the results go back essentially to Hecke (see [84], §7). A deep theorem of Deligne and Serre associates to *any* newform f of weight 1 an irreducible two-dimensional representation with the same L-function (see [84], Theorem 2, p. 305).

The situation in our next example §3.2.3 where $K = \mathbf{Q}(\sqrt{-47})$ with class number $h = 5$ is quite similar to the present one. There are now two newforms F_1, F_2 which we can describe as follows (see [84], §8.1, p. 343). Fix a generator χ of the character group of $Cl(K)$. Then

$$F_j := \tfrac{1}{2} \sum_Q \chi(Q)^j \Theta_Q, \quad j = 1, 2, \tag{3.5}$$

where the sum is over representatives Q of $Cl(K)$ and $\chi(Q)$ denotes the value of χ on the class of Q. Let ρ_1, ρ_2 be the corresponding two-dimensional Galois representations. Then if ρ' denotes the permutation representation of $\mathrm{Gal}(\overline{\mathbf{Q}}/\mathbf{Q})$ on the roots the polynomial f (generating H over K) then

$$\rho' = \rho_1 \oplus \rho_2 \oplus 1.$$

This again explains the relation between the number of fixed points of τ_p and the coefficient a_p of the form F which is visible in the numerical output below.

3.2.3 Dihedral group of order 10

This example is very similar to the previous one so we will be brief. We start with the polynomial $f := x^5 - x^3 + 2x^2 - 2x + 1$. The Galois closure $H/\mathbf{Q}$ of $\mathbf{Q}[x]/(f)$ is an unramified extension of $K := \mathbf{Q}(\sqrt{-47})$ of degree 5. The Galois group G of $H/\mathbf{Q}$ is isomorphic to the dihedral group D_{10} of order 10, symmetries of the regular pentagon in the plane. Since $h(-47) = 5$ we again have that H is the Hilbert class field of K.

Let us start by computing the factorization types of f

```
poltypes(f,2,1000)

[[5], [2, 2, 1], [1, 1, 1, 1, 1]]
```

The conjugacy classes of D_{10} are: $\{1\}, \{\tau, \tau^{-1}\}, \{\tau^2, \tau^{-2}\}, \{\sigma_1, \sigma_2, \sigma_3, \sigma_4, \sigma_5\}$ (compare with (2.13)). In an embedding of $G \hookrightarrow S_5$ the two classes $\{\tau, \tau^{-1}\}$ and $\{\tau^2, \tau^{-2}\}$ both get mapped to the type [5] and the class of the reflections $\{\sigma_1, \sigma_2, \sigma_3, \sigma_4, \sigma_5\}$ gets mapped to the type [2,2,1].

3.2 Examples of reciprocity for imaginary quadratic fields 83

We formulate reciprocity in terms of a weight one newform as we did in §3.2.2. The class group $Cl(K)$ is now cyclic of order 5. We first compute $F = F_1 + F_2$ with F_j given in (3.5).

```
? F=bqftheta([1,-1,12],150)-1/2*bqftheta([2,1,6],150)
    -1/2*bqftheta([3,1,4],150)

2*q - q^2 - q^3 + q^4 - 2*q^6 - q^7 - 2*q^8 + q^9 + 2*q^12
+ 3*q^14 - q^17 + 2*q^18 - 2*q^21 + q^24 + 2*q^25 - 2*q^27
- 3*q^28 + 2*q^32 - 2*q^34 - 2*q^36 - q^37 + q^42 + 2*q^47
+ q^49 - q^50 + 3*q^51 - q^53 + q^54 + q^56 - q^59 - q^61 +
2*q^63 - q^64 + 2*q^68 - q^71 - q^72 - 2*q^74 - q^75 - q^79
+ 4*q^83 - q^84 - q^89 - q^94 - q^96 - q^97 - 3*q^98 +
q^100 - q^101 + q^102 - q^103 + 3*q^106 - q^108 + 3*q^111 +
3*q^118 - 2*q^119 + 2*q^121 + 3*q^122 - q^126 + q^128 -
q^131 + q^136 - q^141 - 2*q^142 + 2*q^147 + 2*q^148 - q^149
+ O(q^150)
```

and then compare with the factorization types.

```
? forprime(p=2,100,print(p, "\t", polcoeff(F,p), "\t",
polfacttype(f,p)))

2       -1      [5]
3       -1      [5]
5       0       [2, 2, 1]
7       -1      [5]
11      0       [2, 2, 1]
13      0       [2, 2, 1]
17      -1      [5]
19      0       [2, 2, 1]
23      0       [2, 2, 1]
29      0       [2, 2, 1]
31      0       [2, 2, 1]
37      -1      [5]
41      0       [2, 2, 1]
43      0       [2, 2, 1]
47      2       []
53      -1      [5]
59      -1      [5]
61      -1      [5]
67      0       [2, 2, 1]
71      -1      [5]
73      0       [2, 2, 1]
79      -1      [5]
83      4       [1, 1, 1, 1, 1]
89      -1      [5]
97      -1      [5]
```

3.2.4 An example of F. Voloch

Let $h := x^2 + x$ and for $n \in \mathbf{N}$ let $h^{(n)}(x) := h \circ h \circ \cdots \circ h(x)$ (h composed with itself n times). Finally let

$$f(x) := (h^{(3)} - x)/x^2 = x^6 + 4x^5 + 8x^4 + 10x^3 + 9x^2 + 6x + 3. \quad (3.6)$$

The goal is to describe the map $p \mapsto \tau_p$ of factorization types of f.

To compute f we use the GP function

```
polit(f,n)=local(g);g=f;for(k=1,n-1,g=subst(g,x,f));g
```

Since the degree of f is only 6 we can use the built-in GP function polgalois to determine its Galois group G.

```
? f=(polit(x^2+x,3)-x)/x^2

x^6+4*x^5+8*x^4+10*x^3+9*x^2+6*x+3;

? polgalois(f)

[18, -1, 1, "F_18(6) = [3^2]2 = 3 wr 2"]
```

As it happens then, G is the unique transitive subgroup of S_6 of order 18 (up to isomorphism). As an abstract group G is isomorphic to the direct product of $\mathbf{Z}/3\mathbf{Z}$ and S_3 (we may easily check this; see the end of this section).

We would like to find a polynomial which will generate the Galois closure $L/\mathbf{Q}$ of $K = \mathbf{Q}[x]/(f)$; we try our function polwedge discussed in §7.2.3, which gives a polynomial whose roots are products of k distinct roots of f.

We are lucky. Taking $k = 3$ we find an irreducible factor of degree 18. Since the field it defines must be contained in L it must equal L.

```
factor(polwedge(f,3)[4])

[x^2 - 2*x + 3 1]

[x^18 + 12*x^17 + 72*x^16 + 276*x^15 + 729*x^14 + 1296*x^13
+ 1116*x^12 - 1782*x^11 - 9558*x^10 - 21276*x^9 - 28674*x^8
- 16038*x^7 + 30132*x^6 + 104976*x^5 + 177147*x^4 +
201204*x^3 + 157464*x^2 + 78732*x + 19683 1]

? g-%[2,1];
```

To make manipulations of this field a little easier we replace g by a simpler version generating the same field using polredabs.

3.2 Examples of reciprocity for imaginary quadratic fields

```
? g=polredabs(g,16)

x^18 + 6*x^16 - 4*x^15 + 15*x^14 - 12*x^13 + 38*x^12 -
54*x^11 + 84*x^10 - 84*x^9 + 84*x^8 - 54*x^7 + 38*x^6 -
12*x^5 + 15*x^4 - 4*x^3 + 6*x^2 + 1
```

We now look for its subfields of degree 3.

```
? sfv=nfsubfields(g,3);
?   length(sfv)
   4
? sfv=vector(4,k,sfv[k][1]);
```

(With the last line we keep in sfv the polynomials defining the cubic fields and throw away their embeddings into L.) Since G has four subgroups of index 3 we find, as we should, four cubic subfields of L. Only one of them, say F, should be Galois (the fixed field of the S_3 factor of G) and indeed

```
? for(k=1,4,print(k," ",polgalois(sfv[k])))
1   [6, -1, 1, "S3"]
2   [6, -1, 1, "S3"]
3   [6, -1, 1, "S3"]
4   [3, 1, 2, "A3"]
```

So we now know that the cyclic cubic field F is defined by

```
? h= sfv[4]

x^3 - 12*x + 8
```

and the discriminant of F is

```
? nfdisc(h)

81
```

Since $F/\mathbf{Q}$ is abelian and only 3 divides its discriminant it must be a subfield of a cyclotomic field $\mathbf{Q}(\zeta_{3^k})$ for some $k \geq 2$ (see the remark at the end of §2.1.2). On the other hand, $(\mathbf{Z}/3^k\mathbf{Z})^\times$ is cyclic of order $\phi(3^k) = 2 \cdot 3^{k-1}$ and hence we can take $k = 2$. Let us check.

```
? nfisincl(h,polcyclo(9))

[-2*x^4 + 2*x^2 - 2*x, -2*x^5 - 2*x^2 + 2*x, 2*x^5 + 2*x^4]
```

It follows that the factorization types of h are described as follows: [1, 1, 1] if $p \equiv \pm 1 \mod 9$; [3] otherwise.

The other three cubic subfields of L are conjugate and have Galois closure a degree 6 field, say $N/\mathbf{Q}$ with Galois group S_3 (the fixed field of the $\mathbf{Z}/3\mathbf{Z}$

factor of G). The polynomial $q := x^3 + 6x + 8$ stored in `sfv[1]` defines a cubic subfield N.

Since the discriminant of f is -41472 and

```
? factor(poldisc(f))

[-1 1]

[2 9]

[3 4]
```

its Galois closure $L/\mathbf{Q}$ contains $K := \mathbf{Q}(\sqrt{-2}) = \mathbf{Q}(\sqrt{-41472})$. From the structure of G we know that this is the only quadratic subfield of $L/\mathbf{Q}$ and hence it must be contained in $N/\mathbf{Q}$. It does not hurt to check this out.

```
? sfv2=nfsubfields(g,2);
? length(sfv2)
1

? sfv2[1][1]
x^2 + 12*x + 54

? nfdisc(%)
-8
```

We would like to identify the abelian extension N/K in terms of reciprocity. In this case, as opposed to the previous two examples, N/K is ramified. The ramified primes are those dividing 3 only: the ramification at 2 in $N/\mathbf{Q}$ is completely absorbed by $K/\mathbf{Q}$.

```
? factor(nfdisc(q)/nfdisc(x^2+2))

[3 4]
```

Hence, the Artin map no longer gives a surjective map from $Cl(K)$ to $Gal(N/K)$ as we had before (in fact we had an isomorphism as we were dealing with the full Hilbert class field of K; in the present case $Cl(K)$ is actually trivial). However, since the extension $N/\mathbf{Q}$ is dihedral ($Gal(K/\mathbf{Q})$ acts on $Gal(N/K) \simeq \mathbf{Z}/3\mathbf{Z}$ by inversion) N/K is what is called a *ring class field* extension (see [26], Chapter II, §9). This means that the Artin map yields an onto map $Cl(\mathcal{O}) \longrightarrow Gal(N/K)$, where $Cl(\mathcal{O})$ is the class group of some order $\mathcal{O}$ in K of conductor 3^k for some $k \geq 1$. In terms of binary quadratic forms this simply amounts to considering the group of classes of primitive forms of discriminant $D = -8 \cdot 3^{2k}$. It is known that in this case (since $Cl(K)$ is trivial) $Cl(\mathcal{O}) \simeq \mathbf{Z}/2 \cdot 3^{k-1}\mathbf{Z}$ (see [26], Chapter II, (7.27)) and therefore we must have $k = 2$.

3.2 Examples of reciprocity for imaginary quadratic fields

We compute

```
? bqf(-8*3^4)

[[1, 0, 162], [2, 0, 81], [9, -6, 19], [9, 6, 19],
 [11, -10, 17], [11, 10, 17]]
```

and $\{(1, 0, 162), (2, 0, 81)\}$ represents the subgroup of order 2 in $Cl(\mathcal{O})$ (see §3.1.4).

The situation is now very similar to that of the previous two examples: the factorization type τ_p of q modulo $p > 3$ is as follows: if $(\frac{-2}{p}) = 1$ then $\tau_p = [1, 1, 1]$ if $Q_p \sim (1, 0, 162)$ or $(2, 0, 81)$ (recall that Q_p is a form $(p, *, *)$ of discriminant $-8 \cdot 3^4$) and $\tau_p = [3]$ otherwise; if $(\frac{-2}{p}) = -1$ then $\tau_p = [2, 1]$.

Finally, we leave it as an exercise for the reader (Ex. 10) to check that the factorization types τ_p of f are given as follows.

If $(\frac{-2}{p}) = 1$:

$$\tau_p = \begin{cases} [1, 1, 1, 1, 1, 1] & \text{if } (i) \text{ and } (ii) \text{ hold} \\ [3, 3] & \text{if only one of } (i) \text{ and } (ii) \text{ hold} \\ [3, 1, 1, 1] & \text{otherwise} \end{cases} \quad (3.7)$$

where to simplify the typesetting we abbreviated

(i) $Q_p \sim (1, 0, 162), (2, 0, 81),$ (ii) $p \equiv \pm 1 \bmod 9.$

If $(\frac{-2}{p}) = -1$:

$$\tau_p = \begin{cases} [2, 2, 2] & \text{if } (ii) \\ [6] & \text{otherwise.} \end{cases} \quad (3.8)$$

Here is a numerical check

```
? forprime(p=5,200,if(kronecker(-2,p)==1, print(p, "\t",
p%9, "\t", bqfred(bqfprimef(p,-8*3^4))[1], "\t",
%polfacttype(f,p))))

11      2       [11, -10, 17]     [3, 1, 1, 1]
17      8       [11, 10, 17]      [3, 3]
19      1       [9, -6, 19]       [3, 3]
41      5       [11, -10, 17]     [3, 1, 1, 1]
43      7       [9, -6, 19]       [3, 1, 1, 1]
```

59	5	[11, 10, 17]	[3, 1, 1, 1]
67	4	[9, 6, 19]	[3, 1, 1, 1]
73	1	[9, 6, 19]	[3, 3]
83	2	[2, 0, 81]	[3, 3]
89	8	[2, 0, 81]	[1, 1, 1, 1, 1, 1]
97	7	[9, -6, 19]	[3, 1, 1, 1]
107	8	[11, 10, 17]	[3, 3]
113	5	[2, 0, 81]	[3, 3]
131	5	[2, 0, 81]	[3, 3]
137	2	[11, -10, 17]	[3, 1, 1, 1]
139	4	[9, 6, 19]	[3, 1, 1, 1]
163	1	[1, 0, 162]	[1, 1, 1, 1, 1, 1]
179	8	[2, 0, 81]	[1, 1, 1, 1, 1, 1]
193	4	[9, -6, 19]	[3, 1, 1, 1]

In fact, we can explicitly compute the Frobenius automorphisms as we did in §2.5.1; GP has no problem computing all the automorphisms of $L/\mathbf{Q}$.

```
? galv=nfgaloisconj(g);
? forprime(p=5,200,if(kronecker(-2,p)==1,print(p, "\t",p%9,
"\t", bqfred(bqfprimef(p,-8*3^4))[1], "\t",
polallfrob(p,galv,g))))
```

11	2	[11, -10, 17]	[7, 7, 7, 10, 10, 10]
17	8	[11, 10, 17]	[6, 6, 6, 16, 16, 16]
19	1	[9, -6, 19]	[6, 6, 6, 16, 16, 16]
41	5	[11, -10, 17]	[8, 8, 8, 11, 11, 11]
43	7	[9, -6, 19]	[7, 7, 7, 10, 10, 10]
59	5	[11, 10, 17]	[8, 8, 8, 11, 11, 11]
67	4	[9, 6, 19]	[8, 8, 8, 11, 11, 11]
73	1	[9, 6, 19]	[6, 6, 6, 16, 16, 16]
83	2	[2, 0, 81]	[14, 14, 14, 14, 14, 14]
89	8	[2, 0, 81]	[1, 1, 1, 1, 1, 1, 1, 1, 1, 1, 1, 1, 1, 1, 1, 1]
97	7	[9, -6, 19]	[7, 7, 7, 10, 10, 10]
107	8	[11, 10, 17]	[6, 6, 6, 16, 16, 16]
113	5	[2, 0, 81]	[5, 5, 5, 5, 5, 5]
131	5	[2, 0, 81]	[5, 5, 5, 5, 5, 5]
137	2	[11, -10, 17]	[7, 7, 7, 10, 10, 10]
139	4	[9, 6, 19]	[8, 8, 8, 11, 11, 11]
163	1	[1, 0, 162]	[1, 1, 1, 1, 1, 1, 1, 1, 1, 1, 1, 1, 1, 1, 1, 1]
179	8	[2, 0, 81]	[1, 1, 1, 1, 1, 1, 1, 1, 1, 1, 1, 1, 1, 1, 1, 1]
193	4	[9, -6, 19]	[8, 8, 8, 11, 11, 11]

We let the reader interpret this output.

3.2 Examples of reciprocity for imaginary quadratic fields

While we are at it, let us check that G is isomorphic to $S_3 \times \mathbf{Z}/3\mathbf{Z}$ as we claimed at the beginning of this section. First we compute the center Z of G. Here are a few GP functions that allow us to do this by brute force; it should be pretty clear how they work (see also Ex. 2.14). (To make them more efficient we should really work modulo a suitable prime ideal, see the remark at the end of §2.5.)

```
galmult(g,h,f)=subst(h,x,g)%f

galiscentral(g,galv,f)=
{
  for(k=1,length(galv),
         if(galmult(g,galv[k],f)!=galmult(galv[k],g,f),
                    return(0)));
  1
}

galcenter(galv,f)=
{
  local(w);
  w=[];

  for(k=1,length(galv),
         if(galiscentral(galv[k],galv,f),w=concat(w,k)));

  w
}
```

Applied to our example we get

```
? galcenter(galv,g)

   [1, 5, 14]
```

We conclude that $Z \simeq \mathbf{Z}/3\mathbf{Z}$ (compare with the output of polallfrob above!)

We now compute the order of the individual elements (as we did in §2.5)

```
? for(k=1,18,print(k,"\t",galorder(galv[k],g)))
```

1	1
2	6
3	2
4	2
5	3
6	3
7	3
8	3

9	6
10	3
11	3
12	6
13	2
14	3
15	6
16	3
17	6
18	6

Let $u \in G \setminus Z$ be an element of order 3. Say

```
? u=galv[6];
```

Multiplication by u permutes the three elements $\sigma_i \in G$ for $i = 1, 2, 3$ of order 2 (entries 3, 4, 13 in galv).

```
? memb(galmult(u,galv[3],g),galv)

 13
? memb(galmult(u,galv[4],g),galv)

 3
? memb(galmult(u,galv[13],g),galv)

 4
```

A little thought will show that $H := \{1, u, u^2, \sigma_1, \sigma_2, \sigma_3\}$ is a subgroup of G isomorphic to S_3 and $G = Z \times H$.

3.2.5 Final comments

The reader might justly be wondering where did the dihedral examples in this chapter come from. This is related to an aspect of reciprocity which we have barely discussed (see remark at the end of §2.1.2): class field theory guarantees the existence of abelian extensions of number fields with prescribed ramification properties. Concretely, for example, for every number field K there is an unramified abelian extension H/K, the *Hilbert class field* of K, for which the Artin map $\Phi : Cl(K) \longrightarrow \text{Gal}(H/K)$ is an isomorphism (see [26], Chapter II, §5 C). In the case that K is an imaginary quadratic field the classical theory of *Complex Multiplication* (see [26], Chapter III) gives a concrete way to generate H using modular functions (see §1.2.3 for a few examples).

3.3 Exercises

1. Check that the routines in this chapter do what was intended. (In particular, why is it not necessary to check the boundary condition $|b| == a$ in the last `if` statement in `bqfred`?)
2. Prove that there are only finitely many reduced forms of a given discriminant (and hence the class number is finite).
3. Find all automorphisms of a positive binary quadratic form $Q = (a, b, c)$.
4. Search for discriminants $D < 0$ such that $h(D) = 1$ (we already encountered the case $D = -4$ in §3.1.2).
5. For a given discriminant D count how many classes C satisfy $C^2 = 1$. Do the calculation theoretically and check your answer numerically in examples. In particular, show that $D = -p$ with $p \equiv 3 \bmod 4$ an odd prime then $Cl(K)$ is odd and conversely.
6. Search for $D < 0$ such that every class C of forms of discriminant D satisfies $C^2 = 1$.
7. Consider positive definite binary Hermitian forms H over $\mathbf{Z}[i]$; i.e.,

$$H(x, y) = a x \bar{x} + b x \bar{y} + \bar{b} \bar{x} y + c y \bar{y},$$

with $a, c \in \mathbf{Z}$ positive, $b \in \mathbf{Z}[i]$ and $D := b\bar{b} - 4ac < 0$. Find an algorithm analogous to that of Gauss to reduce such forms.

8. Use the previous exercise to give an algorithm analogous to that of Cornachia §3.1.2 to find a solution to the equation

$$p = x^2 + y^2 + z^2 + w^2, \qquad x, y, z, w \in \mathbf{Z} \tag{3.9}$$

for any prime p.

9. What is the significance (if any) in whether we get $(2, 1, 4)$ or $(2, -1, 4)$ in the listing after (3.2)?
10. Check that the factorization types of f in §3.2.4 are described as claimed. Interpret the output `polallfrob` given at the end of §3.2.4.
11. Repeat the calculations of §3.2.1 and §3.2.3 for some of the other dihedral examples computed at the end of §1.2.3.

4 Sequences

In this chapter we discuss some computational aspects of sequences of numbers by concentrating on the sequence of trinomial numbers. For well-known sequences we give their label in Sloane's encyclopedia of sequences [86].

4.1 Trinomial numbers

Euler [32] considered the sequence of *trinomial numbers* (or *central trinomial coefficients*), Sloane A002426, defined by

$$t_n := \left[\left(x + 1 + x^{-1}\right)^n\right]_0 = 1, 1, 3, 7, 19, 51, \ldots, \quad n = 0, 1, 2, \ldots$$

where for any Laurent polynomial $P \in \mathbf{C}[x, x^{-1}]$, $[P(x)]_0$ denotes its constant term. This sequence appears often in combinatorial problems.

We will discuss several ways of computing the sequence t_n. For small n we can simply use the definition. For example, in GP we could use the following function.

```
trinomial(n) =
{
  local(pol = 1, pol0 = 1+x+x^2);
  concat(1,vector(n, k, polcoeff(pol*=pol0, k)));
}
```

The output is a vector with the first n trinomial numbers $t_0, t_1, t_2, \ldots, t_n$ (so note that t_k is the $k+1$ entry of the output!)

4.1.1 Formula

By using the multinomial theorem we find that

$$t_n = \sum_{k=0}^{n} \frac{n!}{k!^2 (n-2k)!} \quad (4.1)$$

and this we can program in GP as follows.

```
trinomial1(n) =
{
  local(c = 1);
  1 + sum(k=0, n\2-1, c *= (n-2*k)*(n-2*k-1)/(k+1)^2);
}
```

Note that this function only computes the n-th trinomial number and not the sequence $t_0, t_1, \ldots, t_n$.

4.1.2 Differential equation and linear recurrence

There is a faster way of computing the trinomial numbers; we may exploit the fact that they satisfy a linear recurrence with polynomial coefficients. Assuming this fact for the moment, how do we find the recurrence?

Consider the generating function

$$u = \sum_{n=0}^{\infty} a_n x^n, \tag{4.2}$$

a formal power series in the variable x. A linear recurrence with polynomial coefficients for a_n is equivalent to a linear differential equation with polynomial coefficients for u. Concretely, let L be the differential operator

$$L = \sum_{j=0}^{d}\sum_{k=0}^{r} p_{j,k} x^j \delta^k, \quad p_{j,k} \in \mathbf{C}, \quad \delta = x\frac{d}{dx}. \tag{4.3}$$

Then the n-the coefficient of $L(u)$ is

$$\sum_{j=0}^{d}\sum_{k=0}^{r} p_{j,k}(n-j)^k a_{n-j}$$

where we have set $a_n = 0$ for $n < 0$.

In particular

$$L(u) = 0 \tag{4.4}$$

if and only if

$$\sum_{j=0}^{d}\sum_{k=0}^{r} p_{j,k}(n-j)^k a_{n-j} = 0, \quad n \in \mathbf{Z}. \tag{4.5}$$

This is typically too restrictive. In practice, (4.5) may only hold for $n \geq d$ (i.e., without extending the sequence back with zeros by setting $a_n = 0$ if $n < 0$). We found it convenient to have two functions to deal with each case separately.

Let us consider first the case (4.4). We can find the differential operator L by linear algebra if we have upper bounds for r and d. All we want is a linear relation (with constant coefficients) among $x^j \delta^k u$ with $j = 0, 1, \ldots, d$ and $k = 0, 1, \ldots, r$. A priori, of course, we have no idea how large r and d might be, but we can still search for L, slowly increasing r and d, hoping that we will eventually find it.

In GP we can proceed as follows. First, we consider the following.

```
serlindep(v,l,flag)=
{
  local(M);
  M=matrix(l,length(v),j,k,polcoeff(v[k],j-1));

  M=if(flag,
       qflll(M,4)[1],
       matker(M));
  M[,1]
}
```

Given a vector v of power series in the variable x, given to precision at least $O(x^l)$, this function will return a vector representing a linear combination of the components of v vanishing to order $O(x^l)$.

The argument flag is optional and if non-zero we use the function

```
qflll(*,4)[1]
```

rather than matker to solve the linear system. This works better in the case that our power series has integral coefficients.

Next, we apply serlindep to a vector containing $x^j \delta^k u$, where u is the input series. Here is what the resulting function would look like; note the maneuvering of indices necessary to deal with $x^j \delta^k u$ in a linear fashion.

```
serdiffeq(u,r,d,l,flag = 1)=
{
  local(w,S);
    if(l, u+=O(x^l), l=length(u));
    if(l < (d+1)*(r+1), error("Not enough precision in series."));

  w=vector(r+1); w[1]=u;
    for(k=1,r, w[k+1]=x*deriv(w[k],x));
    v=w; for(j=1,d, v = concat(v, w*=x));

  S=serlindep(v,l,flag);
    if(S, vector(r+1,k, sum(j=0,d, S[j*(r+1)+k,1]*x^j)));
}
```

We added an optional parameter l which chooses how many terms of the input series to use. This is useful since *l* determines the size of the linear system to solve and we might not need as many terms as given by the current precision of *u*. We should point out that `serdiffeq` works equally well (with `flag` set to zero) if the input is a polynomial.

We apply this function to our sequence of trinomial numbers and we find

```
? trins=Ser(trinomial(40));
? serdiffeq(trins,1,1)

? serdiffeq(trins,1,2)

[-3*x^2 - x, -3*x^2 - 2*x + 1]
```

As we said, we slowly increase the bounds and eventually we do find what is the likely answer. To double check, we should run the function with a longer vector of initial data.

```
? trins=Ser(trinomial(100));
? serdiffeq(trins,1,2)

[-3*x^2 - x, -3*x^2 - 2*x + 1]
```

We conclude that most likely $Lu = 0$ with

$$L = 3x^2 + x + (-3x^2 - 2x + 1)\delta. \tag{4.6}$$

We will prove this in the next section §4.1.3.

Let us now turn to the case of the linear recursion where we want

$$\sum_{j=0}^{d}\sum_{k=0}^{r} p_{j,k}(n-j)^k a_{n-j} = 0, \quad n \geq d.$$

We rewrite this as follows

$$Q_d(n)a_{n+d} + Q_{d-1}(n)a_{n+d-1} + \cdots + Q_0(n)a_n = 0, \quad n \geq 0, \tag{4.7}$$

for some polynomials $Q_0, Q_1, \ldots, Q_d$ of degree at most *r*.

The idea for the algorithm is essentially the same, the difference is mostly a question of formatting. We change gears slightly and replace multiplication by *x* by the shift operation

$$s: \quad a_0 + a_1 x + a_2 x^2 + \cdots \mapsto a_1 + a_2 x + a_3 x^2 + \cdots,$$

which we may simply code as

```
u \ x
```

Then
$$\sum_{j=0}^{d}\sum_{k=0}^{r} q_{j,k}\delta^{k}s^{j}\left(\sum_{n\geq 0} a_n x^n\right) = \sum_{n\geq 0}\left(\sum_{j=0}^{d}\sum_{k=0}^{r} q_{j,k}n^k a_{j+n}\right)x^n$$

and $Q_j(x) = \sum_{k=0}^{r} q_{j,k} x^k$.

Here is the corresponding GP function. The ouput is the vector $[Q_0, Q_1, \ldots, Q_d]$. Note the loss of d rows in the linear system of equations if compared to `serdiffeq`.

```
seqlinrec(a,r,d,l,flag = 1)=
{
  local(w,v,S);
    if(!l, l=length(a));
    if(l < (d+1)*(r+1)+d, error("Not enough terms of the sequence."));

    w=vector(d+1); w[1]=Polrev(a);
      for(j=1,d, w[j+1] = w[j]\x);
      v=w; for(k=1,r, v = concat(v, w=x*deriv(w,x)));
      S=serlindep(v,l-d,flag);
        if(S, S=vector(d+1,j, sum(k=0,r, S[k*(d+1)+j]*x^k)));
}
```

Applied to the trinomial example we get

```
?   trinv=trinomial(40);
?   seqlinrec(trinv,1,2)

[-3*x - 3, -2*x - 3, x + 2]

?   subst(%,x,x-2)

[-3*x + 3, -2*x + 1, x]
```

Hence we expect that
$$nt_n + (-2n+1)t_{n-1} + (-3n+3)t_{n-2} = 0, \quad n \in \mathbb{Z}. \quad (4.8)$$

We will prove later §4.1.3 that this in indeed the case. Note that from this point of view, it is rather miraculous that the t_n are actually integers for all n, since this requires that n divides $(-2n+1)t_{n-1} + (-3n+3)t_{n-2}$ for all n, which is not a priori obvious.

4.1 Trinomial numbers

Finally, here is a GP function to compute the trinomial numbers recurrently

```
trinomial2(n) =
{
  local(v);
  v=vector(n+1); v[1] = v[2] = 1;
    for(k=2,n,
      v[k+1] = ((2*k-1)*v[k]+3*(k-1)*v[k-1]) / k);
  v
}
```

To check things out we can do the following

```
? trinomial(500) == trinomial2(500)

1
```

We can easily modify this function so that it only returns the n-th trinomial number; we only need to keep two terms of the sequence at any given time.

```
trinomial3(n) =
{
  local(a, a0 = 0, a1 = 1);
    for(k=1,n,
      a = ((2*k-1)*a1+3*(k-1)*a0) / k;
      a0=a1; a1=a);
  a1
}
```

Finally, here is a GP routine to solve a linear recursion like (4.7) given the initial data $a_0, a_1, \ldots, a_{d-1}$ and N the number of extra terms we would like to compute.

```
recsolve(N,Q,a) =
{
  local(w,q, d = length(Q)-1);
  w = concat(a, vector(N+d+1 - length(a)));
    for(n=0,N,
      q = subst(Q,x,n);
      w[n+d+1] = -sum(k=0,d-1, q[k+1]*w[n+k+1])/q[d+1]);
  w
}
```

For example,

```
? recsolve(10,[-3*x - 3, -2*x - 3, x + 2],[1,1])

[1, 1, 3, 7, 19, 51, 141, 393, 1107, 3139, 8953, 25653, 73789]
```

4.1.3 Algebraic equation

There is yet another way of computing the trinomial numbers: by identifying their generating function as an algebraic function. Let us try this.

```
? u=Ser(trinomial(50));
? 1/u^2

1 - 2*x - 3*x^2 + O(x^50)
```

As suggested by this calculation it is indeed the case that

$$u = \frac{1}{\sqrt{1 - 2x - 3x^2}} \tag{4.9}$$

where u is the generating function (4.2). Hence, the following function will also compute the first $n+1$ trinomial numbers.

```
trinomial4(n) = Vec(1/sqrt(1 - 2*x - 3*x^2+O(x^(n+1))))
```

Here is a sketch of the proof of (4.9). The constant term of a Laurent polynomial $P(x) \in \mathbb{C}[x, x^{-1}]$ can be expressed as the integral

$$[P]_0 = \frac{1}{2\pi i} \int_{|z|=1} P(z) \frac{dz}{z}. \tag{4.10}$$

Consequently, we have

$$u = \sum_{n=0}^{\infty} \frac{1}{2\pi i} \int_{|z|=1} P^n(z) \frac{dz}{z} x^n \tag{4.11}$$

and by summing the geometric series

$$u = \frac{1}{2\pi i} \int_{|z|=1} \frac{1}{1 - xP(z)} \frac{dz}{z}. \tag{4.12}$$

In our case $P(z) = z + 1 + 1/z$ and the integrand is $1/Q(x,z)$, where

$$Q(x,z) = -xz^2 + (1-x)z - x. \tag{4.13}$$

The discriminant of Q in the variable z is none other than $1 - 2x - 3x^2 = (1+x)(1-3x)$. It is enough to apply the residue theorem to finish the proof; we leave the details to the reader. It is now also easy to check that $Lu = 0$ with L as in (4.6) hence proving the linear recursion (4.8).

In general, if we have a power series u to some precision we can search for an algebraic equation that it might satisfy in a similar way to how we searched for a differential equation. Concretely, we may use the following.

```
seralgdep(u,r,d,l,flag = 1)=
{
  local(w,v);

  if(l, u+=O(x^l), l=length(u));
  if(l < (d+1)*(r+1), error("Not enough precision on input."));

  w=vector(d+1); w[1]=1;
  for(k=1,d, w[k+1] = u*w[k]);
  v=vector((d+1)*(r+1),k, w[(k-1)\(r+1)+1] * x^((k-1)%(r+1)));
  v=serlindep(v,l,flag);
    if(v,
      v/=content(v);
      v=vector(d+1,k, sum(j=0,r,v[(k-1)*(r+1)+j+1,1]*x^j));
      sum(k=0,d,v[k+1]*y^k));
}
```

The output, if non-empty, is a polynomial $A(x,y)$ satisfying (as far as the precision goes) $A(u(x),x) = 0$.

Let us test in our running example

```
? u=Ser(trinomial(30));
? seralgdep(u,2,2)

3*y^2*x^2 + 2*y^2*x + (-y^2 + 1)
```

Remark In practice when using the algorithms of this section serdiffeq, serlinrec, seralgdep, if the coefficients are integers, it is convenient to reduce the input modulo a suitable number m, which will typically cut the running time significantly. If the algorithms produce no answer we know that we need to increase the parameters in the input. If we obtain an answer modulo m we can run it again with the original input to find an integral answer (or use recognizemod of §1.2.2 to attempt to recover it from the solution modulo m).

4.1.4 Hensel's lemma and Newton's method

We will now describe how Hensel's lemma and its generalizations can be used to compute the coefficients in the power series expansion of an algebraic function about a regular point. However, as Comtet [25] pointed out, any such function satisfies a linear differential equation with polynomial coefficients. Hence its Taylor coefficients satisfy a linear recursion also with polynomial coefficients, which typically gives the best way to compute them. See §4.2.4 for a simple example.

For the trinomial example the algebraic equation is

$$f(x, u) = u^2(-3x^2 - 2x + 1) - 1 = 0 \qquad (4.14)$$

and we want a power series solution about $x = 0$.

Suppose that we already have polynomials w_k for $k = 0, 1, \ldots, n-1$, of degree $\deg(w_k) = k$, and satisfying

$$f(x, w_k) = O(x^{k+1}), \qquad w_k = w_{k+1} + O(x^{k+1}), \quad 0 \le k < n.$$

Set $w_n = w_{n-1} + zx^n$ with z to be determined. By Taylor's theorem

$$f(x, w_n) = f(x, w_{n-1}) + \frac{\partial f}{\partial u}(x, w_{n-1}) z x^n + O(x^{n+1}) \qquad (4.15)$$

By hypothesis, $f(x, w_{n-1}) = x^n h$ for some polynomial h. Hence, to solve $f(x, w_n) = O(x^{n+1})$ we divide (4.15) by x^n and get

$$0 = h + \frac{\partial f}{\partial u}(0, w_0) z + O(x) \qquad (4.16)$$

(note that $\frac{\partial f}{\partial u}(x, w_{n-1}) = \frac{\partial f}{\partial u}(0, w_0) + O(x)$ since $w_k = w_0 + O(x)$ for all k).

Therefore, we can solve for z (and find w_n) if $\frac{\partial f}{\partial u}(0, w_0) \ne 0$. This is the simplest form of Hensel's lemma for power series (see §4.2.4 for the general case). The same exact argument works in the p-adic setting (see §6.1.2)

Here is a GP implementation.

```
hensel(f,w,n) =
{
  local(g,d,c);
  d=poldegree(w);
  c=-1/truncate(subst(subst(deriv(f,u),x,0),u,w+O(x)));
    for(k=d+1,d+n,
      g=subst(f,u,w+O(x^(k+1)));
      g=polcoeff(g,k);
      w+=c*g*x^k);
  w+O(x^(d+n+1))
}
```

Applied to the trinomial equation (4.14) we get

```
? f=u^2*(-3*x^2-2*x+1)-1;
? hensel(f,1,5)

1 + x + 3*x^2 + 7*x^3 + 19*x^4 + 51*x^5 + O(x^6)
```

We actually can do things more efficiently, closer to the original method of Newton for solving algebraic equations.

Suppose we have a rational function $r(x)$ which is a solution of $f(x, u) = 0$ to order n; i.e., suppose that

$$f(x, r) = x^n h(x) \tag{4.17}$$

for some rational function $h(x)$ with $v_x(h) = 0$, where v_x is the the valuation at $x = 0$ (order of zero or pole at $x = 0$). Taylor's theorem then implies that

$$0 = f(x, r + zx^n) \equiv hx^n + \frac{\partial f}{\partial u}(x, r)zx^n \mod x^{2n}. \tag{4.18}$$

If we assume

$$v_x\left(\frac{\partial f}{\partial u}(x, r)\right) = 0 \tag{4.19}$$

then we can, as before, divide through by x^n and find

$$z = \frac{h}{\frac{\partial f}{\partial u}(x, r)} \mod x^n. \tag{4.20}$$

Hence we may take as a new approximation

$$r - \frac{f(x, r)}{\frac{\partial f}{\partial u}(x, r)}. \tag{4.21}$$

This new rational function is an approximate solution to order $O(x^{2n})$. It is easy to check that the condition (4.19) remains valid for this new approximation and we may therefore iterate this process doubling the precision at each step.

The resulting iterative algorithm has a particularly pleasing form in the case of square roots, i.e., for equations of the form $f(u, x) = u^2 - D(x)$ for, say, a polynomial $D(x)$ with $D(0) = 1$. Namely,

$$r_{n+1} = \tfrac{1}{2}\left(r_n + \frac{D}{r_n}\right), \quad r_0 = 1. \tag{4.22}$$

Note that $\pm\sqrt{D}$ are the only fixed points of the map $x \mapsto \tfrac{1}{2}(x + D/x)$.

If we write $r_n = A_n/B_n$ (not necessarily in lowest terms) we can homogenize the iteration in the form

$$(A_{n+1}, B_{n+1}) = (A_n^2 + B_n^2 D, 2A_n B_n), \quad A_0 = B_0 = 1. \tag{4.23}$$

4 : Sequences

Here is an implementation of this algorithm in GP.

```
newtsqrt(D, n = 5, a = 1, b = 1) =
{
  local(aux);
    for(k=1,n,
      aux=a^2+D*b^2;
      b*=2*a;
      a=aux);
  a/b
}
```

The inputs are: D, what we want the square root of; a, b an initial approximation with default a = b = 1; n the number of iterations wanted, with default of 5. (Note that this algorithm with no changes will also compute an approximation to the square root of a positive real number D. See also §6.1.2 for the p-adic case.)

Notice that though we do double the precision in each step the approximation is not as efficient as it could be: r_n is of degree at most $m := 2^n$ but approximates the square root only to order $O(x^m)$. On the other hand, for any m there does exist a rational function of degree at most m which approximates the square root to order $O(x^{2m+1})$ (the so-called Padé approximation, see the end of next section §4.1.5).

Here is what we get in the trinomial case:

```
? r=1/newtsqrt(1-2*x-3*x^2,3)

(-216*x^6 - 432*x^5 + 432*x^4 + 896*x^3 - 256*x^2 - 384*x +
128)/(81*x^8 + 216*x^7 - 648*x^6 - 1632*x^5 + 304*x^4 + 1664*x^3 -
128*x^2 - 512*x + 128)

? r+O(x^9)

1 + x + 3*x^2 + 7*x^3 + 19*x^4 + 51*x^5 + 141*x^6 + 393*x^7 +
141695/128*x^8 + O(x^9)
```

As pointed out by K. Belabas, we do not really need to do the iteration with rational functions, we may just use their approximations modulo x^{2^n} instead.

```
newtsqrt1(D, n = 5, a = 1) =
{
  local(m);
  m=1;
  for(k=1,n, m = shift(m,1); a = truncate((D+O(x^m))/a + a)/2);
  a+O(x^m)
}
```

This gives the answer as a power series and is significantly faster than `newtsqrt`. Compare the following calculation with our previous one

```
? r=1/serrecognize(newtsqrt1(1-2*x-3*x^2,3))

(-2*x^4 - 4*x^3 + x^2 + 10*x - 4)/(-4*x^4 - 17*x^3 - x^2 +
14*x - 4)

? r+O(x^9)

1 + x + 3*x^2 + 7*x^3 + 19*x^4 + 51*x^5 + 141*x^6 +
393*x^7 + 2209/2*x^8 + O(x^9)
```

(for a description of `serrecognize` see §4.2.3).

See [54] and [98] for a full treatment of the above questions and [33] for a comparison between the linear algorithm (Hensel) versus the quadratic (Newton).

4.1.5 Continued fractions

For $z = a_{-N}x^{-N} + a_{-N+1}x^{1-N} + \cdots a_{-1}x^{-1} + a_0 + a_1x + \cdots$ a Laurent series in x we define its continued fraction expansion $[c_0, c_1, \ldots]$ recursively as follows. We start with $z_0 = z$ and having found z_n we let c_n be the polynomial in x^{-1} such that $v_x(z_n - c_n) \geq 1$ and set $z_{n+1} = (z_n - c_n)^{-1}$. For example, $c_0 = a_{-N}x^{-N} + a_{-N+1}x^{1-N} + \cdots a_{-1}x^{-1} + a_0$. In GP this recursion would look something like this.

```
sercontfrac(z) =
{
  local(v,c);
  c=truncate(z+O(x));
  v=[c];
  z-=c;
  while(z,
          z=1/z;
          c=truncate(z+O(x));
          z-=c;
          v=concat(v,c));
  v
}
```

Let

$$w = \frac{x}{\sqrt{1+bx+ax^2}} = x - \tfrac{1}{2}bx^2 + (\tfrac{3}{8}b^2 - \tfrac{1}{2}a)x^3 + \cdots. \quad (4.24)$$

4 : Sequences

Computing its continued fraction with sercontfrac we find that $c_0 = 0, c_1 = \frac{1}{2}L$ and apparently for $n \geq 2$

$$c_n = \begin{cases} -\frac{4L}{D} & \text{if } n \text{ is even,} \\ L & \text{if } n \text{ is odd,} \end{cases} \qquad (4.25)$$

where $L := b + 2x^{-1}$ and $D := b^2 - 4a$. We will verify this shortly.

Define polynomials in x^{-1} $p_n(x^{-1}), q_n(x^{-1})$ by

$$\begin{array}{lll} p_{-2} = 0, & p_{-1} = 1, & p_n = c_n p_{n-1} + p_{n-2} \\ q_{-2} = 1, & q_{-1} = 0, & q_n = c_n q_{n-1} + q_{n-2}. \end{array} \qquad (4.26)$$

Then p_n/q_n is the rational function with continued fraction expansion $[c_0, c_1, \ldots, c_n]$, a good approximation to $z = [c_0, c_1, \ldots]$. The following formulas are classical (and apply to continued fractions of both real numbers and Laurent series), see for example [2], §5, though note that their x is our x^{-1} (see warning below).

$$q_n p_{n-1} - p_n q_{n-1} = (-1)^n, \quad (n \geq -1) \qquad (4.27)$$

$$z = \frac{p_n z_{n+1} + p_{n-1}}{q_n z_{n+1} + q_{n-1}} = [c_0, c_1, \ldots, c_n, z_{n+1}] \quad (n \geq -1) \qquad (4.28)$$

$$z_n = [c_n, c_{n+1}, \ldots], \quad (n \geq 0) \qquad (4.29)$$

$$q_n z - p_n = (-1)^n/(q_n z_{n+1} + q_{n-1}), \quad (n \geq -1). \qquad (4.30)$$

We see that $\{c_n\}$ is eventually periodic if and only if $\{z_n\}$ is. If $\{z_n\}$ has period N we have

$$z_n = [c_n, c_{n+1}, \ldots, c_{n+N-1}, z_n] \qquad (4.31)$$

for all sufficiently large n, and hence also,

$$z_n = \frac{Az_n + B}{Cz_n + D}, \qquad (4.32)$$

for some polynomials A, B, C, D (independent of n). In turn this means that z_n satisfies a quadratic equation (concretely, $Cz_n^2 + (D-A)z_n - B = 0$) and therefore so does z itself.

WARNING: Usually (for example [2]) the above discussion is applied to Laurent series in x^{-1}; i.e., series of the form $a_N x^N + a_{N-1} x^{N-1} + \cdots a_1 x^1 +$

$a_0 + a_{-1}x^{-1} + \cdots$. (We made our choice in order to be able to use the power series capability of GP.) In particular, sercontfrac will not agree with GP built-in contfrac applied to a rational function. Here is a simple example to illustrate the difference

```
? contfrac((x^3+x-1)/(x^5+x^4+x+1))

[0, x^2 + x - 1, 1/3*x^2 + 1/3, -3*x]
? sercontfrac((x^3+x-1)/(x^5+x^4+x+1))

[-1, (x + 1)/(2*x), (8*x - 4)/x, (-5*x - 3)/(54*x), (-135*x
+ 81)/x, (2*x + 1)/(243*x)]

? sercontfrac(subst((x^3+x-1)/(x^5+x^4+x+1),x,1/x))

[0, (-x^2 + x + 1)/x^2, (x^2 + 1)/(3*x^2), -3/x]
```

There is a built-in function in GP contfracpnqn which will find p_n and q_n given $[c_0, c_1, \ldots, c_n]$; in fact, it will return the matrix

$$\begin{pmatrix} p_n & p_{n-1} \\ q_n & q_{n-1} \end{pmatrix}.$$

For example, for c_n as in (4.25) we have

```
? m=contfracpnqn([0,a/2+1/x,z]);m[1,1]/m[2,1]

2*z*x/((z*a + 2)*x + 2*z)
```

In other words,

$$z_0 = \frac{2xz_2}{(bx+2)z_2 + 2x}; \tag{4.33}$$

inverting this relation between z_0 and z_2 we find

$$z_2 = \frac{-2xz_0}{(bx+2)z_0 - 2x}. \tag{4.34}$$

The continued fraction expansion of z_2 is the purely periodic one $[c_2, c_3, c_2, c_3, \ldots]$ and therefore satisfies the quadratic equation $z_2 = [c_2, c_3, z_2]$ from which we may recover the equation satisfied by z_0 using (4.33). In GP we can do it as follows.

WARNING: In the present context of Laurent series (as opposed to that of real numbers) the converse, i.e, that if z satisfies a quadratic equation then its continued fraction is periodic, is not true in general; it is, in fact, pretty rare (see [2], §5 and Ex. 6).

```
? z2=-2*x*z/((b*x+2)*z-2*x);
? m=contfracpnqn([(-4*b*x - 8)/((b^2 - 4*a)*x), (b*x +
2)/x,z2]);

? p=numerator(m[1,1]/m[2,1]-z2)

16*z^2*a - 16)*b*x^3 + (16*z^2*b^2 + (32*z^2*a - 32))*x^2
+48*z^2*b*x+ 32*z^2
```

This is the quadratic equation satisfied by our continued fraction. It has no linear term in z and

```
? -polcoeff(p,2,z)/polcoeff(p,0,z)

(a*x^2 + b*x + 1)/x^2
```

Hence we have verified that indeed the continued fraction $z = [c_0, c_1, \ldots]$ with the coefficients as in (4.25) satisfies the same algebraic equation of degree two as (4.24) and since they agree in the first few terms in their Laurent expansion they must be equal. In other words, we have now proved that our numerical observation (4.25) holds for all $n \geq 2$.

To return to the sequence of trinomial numbers, we can see now how to write a good rational approximation to their generating function using the periodicity of the continued fraction expansion of (4.24). For c_n that are ultimately periodic rather than using contfracpnqn it is a good idea to write a tailor-made script:

```
contfracper(ci,cp, m = 1)=
{
  local(v,v0,v1);
  v0=[0,1];v1=[1,0];
    for(n=1,length(ci),
      v=ci[n]*v1+v0;
      v0=v1;v1=v);
    for(k=1,m,
    for(n=1,length(cp),
      v=cp[n]*v1+v0;
      v0=v1;v1=v));
  v1[1]/v1[2]
}
```

The input for this function consists of two vectors ci and cp with the initial and periodic part of the coefficients c_j respectively, and an index m, with default of 1, for the number of iterations of the periodic part wanted.

4.1 Trinomial numbers

For the trinomial case we take $a = -3, b = -2$ so that $c_1 = x^{-1} - 1, c_2 = \frac{1}{2}(x^{-1} - 1), c_3 = 2(x^{-1} - 1)$. We compute then

```
? L=-2+2/x;contfracper([0,L/2],[-L/4,L],2)

(-x^5 + 2*x^4 + 3*x^3 - 4*x^2 + x)/(-x^5 - 5*x^4 + 5*x^3 +
5*x^2 -   5*x + 1)

? %+O(x^12)

x + x^2 + 3*x^3 + 7*x^4 + 19*x^5 + 51*x^6 + 141*x^7 +
393*x^8 + 1107*x^9 + 3139*x^10 + 8951*x^11 + O(x^12)
```

Note how the coefficients agree with the trinomial numbers up to the coefficient of x^{10}. This is of course what we expect (Ex. 5): the approximation p_n/q_n to z given by truncating the continued fraction expansion is a Padé approximant. I.e. $q_n z = p_n + O(x^m)$ with $m = \deg(p_n) + \deg(q_n) + 1$. Compare this with the remark at the end of the previous section §4.1.4.

Finally, we can actually also run these routines with variables. For example, take $a = 1, b = -2t$; then the n-th coefficient in the power series expansion of $1/\sqrt{1 - 2tx + x^2}$ is the n-th *Legendre polynomials* $P_n(t)$.

```
? L=-2*t+2/x;contfracper([0,L/2],[-1/(t^2-1)*L,L],1)

((-3*t^2 - 1)*x^3 + 8*t*x^2 - 4*x)/((t^3 + 3*t)*x^3 +
(-9*t^2 - 3)*x^2 + 12*t*x - 4)

? %+O(x^7)

x + t*x^2 + (3/2*t^2 - 1/2)*x^3 + (5/2*t^3 - 3/2*t)*x^4 +
(35/8*t^4 - 15/4*t^2 + 3/8)*x^5 + (63/8*t^5 - 35/4*t^3 +
15/8*t)*x^6 + O(x^7)
```

and this checks

```
? pollegendre(5,t)

63/8*t^5 - 35/4*t^3 + 15/8*t
```

4.1.6 Asymptotics

We can also give an asymptotic expansion for t_n of the form

$$t_n \sim \kappa^n n^{-\alpha} c(1 + c_1 n^{-1} + c_2 n^{-2} + \cdots), \quad n \to \infty. \tag{4.35}$$

Taking a few terms of this series gives a good approximation of t_n for large n. The series itself diverges however, and hence, for a fixed n, we cannot increase the accuracy by adding more terms indefinitely.

See [70] for more on asymptotics of sequences. The most common example is, of course, *Stirling's approximation*

$$n! \sim \sqrt{2\pi n}\left(\frac{n}{e}\right)^n, \quad n \to \infty \tag{4.36}$$

and, more generally, for $x \to \infty$

$$\log \Gamma(x) \sim (x - \tfrac{1}{2})\log x - x + \tfrac{1}{2}\log 2\pi + \sum_{n \geq 1} \frac{B_{2n}}{2n(2n-1)} x^{1-2n}, \tag{4.37}$$

where Γ is the classical Gamma function (see (6.88)) and the B_{2n} are the even-indexed Bernoulli numbers (see §1.3).

A proof of the existence of (4.35) relies on Laplace's method; we give a brief sketch. First note (Ex. 2) that

$$\frac{1}{\pi}\int_{-1}^{1} t^{2n} \frac{dt}{\sqrt{1-t^2}} = \frac{1}{4^n}\binom{2n}{n}. \tag{4.38}$$

Hence

$$t_n = \frac{1}{\pi}\int_{-1}^{1} (1+2t)^n \frac{dt}{\sqrt{1-t^2}}. \tag{4.39}$$

The basic idea is now that for large n the main contribution to this integral comes from the neighborhood of $t = 1$ where $1 + 2t$ is the largest. Making the change of variables $t = 1 - 3/2u$ we obtain

$$t_n = \frac{3^{n+1/2}}{2\pi}\int_0^{4/3} (1-u)^n \frac{du}{\sqrt{u(1-\tfrac{3}{4}u)}}$$

where now the relevant point is $u = 0$. The upper limit $u = 4/3$ is irrelevant for the asymptotics and we will in fact truncate the integral to the interval $[0, 1]$. This truncation only changes the integral by a lower order exponential term which we may ignore. We then make the change of variables $u = 1 - e^{-v}$ to get

$$t_n \sim \frac{3^{n+1/2}}{2\pi}\int_0^{\infty} e^{-nv} f(v) dv, \tag{4.40}$$

where

$$f(v) = \frac{2e^{-v}}{\sqrt{(1-e^{-v})(1+3e^{-v})}}.$$

We may expand f as a series

$$f(v) = v^{-1/2}\sum_{k=0}^{\infty} d_k v^k$$

which when combined with (4.40) gives

$$t_n \sim \frac{3^{n+1/2}}{2\pi} \sum_{k=0}^{\infty} \frac{\Gamma(k+1/2)}{n^{k+1/2}} d_k$$

and therefore

$$\kappa = 3, \quad \alpha = 1/2, \quad c = \sqrt{\frac{3}{4\pi}}, \quad c_k = \frac{(2k)! d_k}{4^k k!}. \tag{4.41}$$

We can compute the first few of the coefficients d_k as follows.

```
? 2*exp(-x)/sqrt((1-exp(-x))/x*(1+3*exp(-x)))+O(x^5)

1 - 3/8*x + 1/384*x^2 + 9/1024*x^3 + 319/163840*x^4 + O(x^5)
```

Hence
$$d_0 = 1, \quad d_1 = -3/8, \quad d_2 = 1/384, \ldots$$
and
$$c_1 = -3/16, \quad c_2 = 1/512, \quad c_3 = 135/8192 \ldots.$$

We can test the asymptotics numerically, for example,

```
? trinomial1(500)

79420966071292168004033106044518546463337896474938505
45779845905518277865240517956805865396862536593984010
84167174314882153268491052760509865516440497866157533
7 97283088773432874906247710933490967467548579137688 9
5835125583280707030468936 9

? %/3.^500

0.0218427746722080409424573901 5
```

agrees reasonably well with

```
? sqrt(3/4/Pi)/sqrt(500)
0.02185096861184158141086771411
```

which we obtain with one term in the asymptotic expansion.

4.1.7 More coefficients in the asymptotic expansion

If we want a lot more coefficients c_k of the asymptotic expansion of t_n, using the defining power series of d_k as we used in the previous subsection is not very efficient. We can do better by directly solving the linear recursion (4.8) in formal series of the form

$$\kappa^n n^{-\alpha}(c_0 + c_1 n^{-1} + c_2 n^{-2} + \cdots). \tag{4.42}$$

Plugging in (4.42) in (4.8) and considering the coefficient of $n^{1-\alpha}$, the smallest possible power of $1/n$, in the result we conclude that we must have

$$\kappa^2 - 2\kappa - 3 = 0.$$

Of the two solutions $\kappa = -1$ and $\kappa = 3$, we know by (4.41) that t_n corresponds to the second. To simplify matters let us renormalize and consider the sequence $u_n := t_n/3^n$, which satisfies the linear recurrence

$$3n u_n + (-2n + 1)u_{n-1} + (-n + 1)u_{n-2} = 0. \tag{4.43}$$

Considering the coefficient of $n^{-\alpha}$ we confirm what we found before (4.41) that $\alpha = 1/2$.

Let

$$C(z) = 1 + \sum_{k=1}^{\infty} c_k z^k. \tag{4.44}$$

By letting $z = n^{-1}$ so that $(n-k)^{-1} = z/(1-kz)$ we can write (4.43) as the following equation in power series in z

$$0 = 3C(z) + \frac{(-2+z)}{\sqrt{1-z}} C\left(\frac{z}{1-z}\right) + \frac{(-1+z)}{\sqrt{1-2z}} C\left(\frac{z}{1-2z}\right), \tag{4.45}$$

where $\sqrt{1-z}$ is the standard branch with value 1 at $z = 0$. This determines C uniquely up to a multiplicative constant and allows us to compute the ratios c_k/c_0 recursively. In GP this can be done as follows.

```
trinasympt(n) =
{
  local(A,B,g,c);
  c=1;
  A=(-2+x)/sqrt(1-x+O(x^(n+2)));
  B=(-1+x)/sqrt(1-2*x+O(x^(n+2)));
    for(k=1,n,
      g=c+z*x^k+O(x^(k+2));
      g=3*g+A*subst(g,x,x/(1-x))+B*subst(g,x,x/(1-2*x));
      g=polcoeff(g,valuation(g,x));
      c=c-polcoeff(g,0)/polcoeff(g,1)*x^k);
  c+O(x^(n+1))
}
```

For example,

```
? trinasympt(5)

1 - 3/16*x + 1/512*x^2 + 135/8192*x^3 + 6699/524288*x^4 -
26397/8388608*x^5 + O(x^6)
```

4.1.8 Can we sum the asymptotic series?

The simple answer is no. The power series $C(z)$ only converges at $z = 0$. For any fixed, small enough z the terms $c_k z^k$ in $C(z)$ will roughly decrease in size up to some point and then start to grow very fast (for large k we have $|c_k| \sim \alpha k! \beta^k$ for some positive constants α, β). For example, for $z = 1$ we see that the terms start fairly small, oscillate and then grow steadily after the 11-th term.

```
? C=trinasympt(20);
? for(k=0,20,print(k,"   ",abs(polcoeff(C,k))*1.))

0    1.0000000000000000000000000000
1    0.18750000000000000000000000000
2    0.0019531250000000000000000000000
3    0.016479492187500000000000000000
4    0.012777328491210937500000000000
5    0.0031467676162719726562500000000
6    0.028084855526685714721679687500
7    0.031100489431992173194885253910
8    0.063062871782676666043698787690
9    0.27148260839919657883001491430
10   0.091006072791170566915752715430
11   2.0590623683524922338250462420
12   5.5462915100390721735712507010
13   11.245505087991235652683172370
14   108.28135752640904113822079920
15   120.19006529997588708601550580
16   1699.5047426921970727361153710
17   8081.7208722254989976416443410
18   14296.743425515019372704241800
19   275527.69203749379131077002800
20   605794.20405314532625840846670
```

which is the fairly typical behavior of asymptotic series.

Despite the inconvenient fact that the series diverges we may still hope to sum just enough terms of the series to obtain an approximation to what its sum should be. Without getting into the subtle issues involved let us try the following: truncate the series at the term of smallest absolute value.

4 : Sequences

Here is one way to do this. (We assume the input power series C has non-zero constant term.)

```
sumasympt(C,z) =
{
  local(w,n);
  w=vector(length(C),k, abs(polcoeff(C,k-1)) * z^(k-1));
  n=vecsort(w,,1)[1];
  [ subst(truncate(C+O(x^n)),x,z), n ]
}
```

We also output the index where we truncated to see if we could have used more terms of C.

We can then define an approximation to $t_n/3^n$ as follows

```
trinapprox(n) =
{
  local(s);
  s=sumasympt(C,1./n);
  [1/sqrt(n)*sqrt(3/4/Pi)*s[1],s[2]]
}
```

Let us test it.

```
? \ps 102
    seriesprecision = 102 significant terms
? C=trinasympt(100);

? trinv=trinomial(25);for(n=1,25,zz=trinapprox(n); print(n,
"\t", zz[2], "\t", abs(zz[1]-trinv[n+1]/3.^n)))

1       3       0.064610509368849493427452580l9
2       11      0.019043483752932649l4417623509
3       11      0.005471007209965012191745080863
4       11      0.001615725942636635563803401664
5       19      0.000488731883753669252967603418 2
6       29      0.000150175008494754354923501619 3
7       29      0.0000466727308585317111979417899 5
8       29      0.0000146305311385104255163550096 8
9       29      0.00000461l42151037138237742700635
10      29      0.0000014649692946100335054860705l5
11      34      0.00000046688079833432202410469469 87
12      34      0.00000014934455216433318515197568 31
13      57      0.000000047921942972579965151973971 7
14      57      0.000000015418731377865015040765515 09
15      57      0.0000000049725331480091631621247615 04
16      57      0.0000000016069258974936148747134893 73
17      57      0.00000000052023481864426014943476899 08
```

18	57	1.686948445341175761636518387 E-10
19	57	5.478100855849705825351910092 E-11
20	57	1.7812378539447477955 E-11
21	85	5.798613508771342358 E-12
22	85	1.8896904310792320548 E-12
23	85	6.164264897739723100 E-13
24	85	2.0126159577007831532 E-13
25	85	6.576556749121997173 E-14

Not a bad fit.

Finally, let us take $n = 500$.

```
? trinapprox(500)

[0.02184277467220804094245739015, 101]
```

which agrees with the exact value

```
0.02184277467220804094245739015
```

we computed at the end of (4.1.6) to 28 decimal places!

4.2 Recognizing sequences

Suppose you somehow produce a sequence $a_0, a_1, \cdots$ of integers. You may want to know if this sequence has appeared in another context (and thereby hopefully gaining insight into the process that generated your sequence) or simply if it can be described in some easy way. For a general sequence the best bet is to use Sloane's on-line encyclopedia of sequences [86] (an invaluable source for identifying and learning about sequences). After typing in the first few terms of your sequence the underlying search engine will produce a matching with the numerous sequences it "knows", if there is one. Sometimes, however, the sequence has a very simple form, for example $a_n = p(n)$ for some polynomial $p(x)$, and the encyclopedia is not the place to look for it.

4.2.1 Values of a polynomial

The problem of recognizing if a sequence a_n of complex numbers is given by the values of a polynomial is quite simple. All we need to do is compute the successive differences $\Delta^k a_n$, where $\Delta a_n := a_{n+1} - a_n$ for $k = 1, 2, \ldots$. The sequence is given by the values of a polynomial of degree d if and only if $\Delta^{d+1} a_n$ is identically zero. We cannot, of course, check these infinitely many conditions but we can still search for a candidate polynomial. We can implement this idea in GP as follows.

4: Sequences

```
seqispol(v,d) =
{
  local(p,l);
  p=v[1]; l=length(v);
    for (k = 1, d+1,
      for(j=1,l-1, v[j]=v[j+1]-v[j]); v[l] = 0;
        if (!v, return(p));
      l--; p += binomial(x,k)*v[1]);
}
```

The input for this function is a vector v containing the sequence to test, d a bound on the degree of the polynomial we are looking for. and m. The output is a polynomial p of degree at most d satisfying $p(n+m) = v[n+1]$ for $n = 0, 1, 2, \ldots$ if such a polynomial exists or 0 otherwise.

We exploited the fact that we know the values of p at consecutive integers. If we only know the values at arbitrary points we should use Lagrange interpolation instead. This can be done with the built-in GP function polinterpolate that will give the unique polynomial p of degree at most $N - 1$ satisfying $p(x_n) = y_n$ for two given sequences $x_1, \ldots, x_N$ and $y_1, \ldots, y_N$ of complex numbers (with $x_i \neq x_j$ for $i \neq j$). We can use this function to test our sequence by taking $x_n = n, y_n = a_n$ for $n = 0, 1, 2, \ldots, N$ and successively increasing N; if the resulting polynomial stabilizes after a while, that is the likely answer.

4.2.2 Values of a rational function

How do we recognize if a sequence is given by the values of a rational function? If we pose that

$$a_n = \frac{p(n)}{q(n)}, \quad n = 0, 1, 2, \ldots \qquad (4.46)$$

for certain polynomials $p(x) = \sum_{k=0}^{d} p_k x^k$ and $q(x) = \sum_{j=0}^{d} q_j x^j$ then (4.46) is equivalent to the following linear system of equations in the indeterminates $p_0, p_1, \ldots, p_d, q_0, q_1, \ldots, q_d$

$$\sum_{j=0}^{d} a_n n^j q_j - \sum_{k=0}^{d} n^k p_k = 0, \quad n = 0, 1, 2, \ldots \qquad (4.47)$$

We can truncate this infinite set of equations by considering only $n = 0, 1, \ldots, N$ for large enough N and then solve the resulting system. As before, we either find a likely candidate for the rational function $r(x) = p(x)/q(x)$ we are after or we prove that no such function exists with $\deg p, \deg q \leq d$, in which case we may want to repeat the process with a larger value of d.

4.2 Recognizing sequences

As with Lagrange interpolation the particular linear algebra problem we generated (4.47) can be solved more efficiently with specific methods (see for example [30]) but we have not pursued this.

Here is an implementation in GP of this idea. As with serlindep with added an optional flag variable for the case of rational entries. In that case we consider the system

$$\sum_{j=0}^{d} r_n n^j q_j - \sum_{k=0}^{d} s_n n^k p_k = 0, \quad n = 0, 1, 2, \ldots \quad (4.48)$$

instead, where $a_n = r_n/s_n$ and use qflll to find a solution.

```
seqisratnl(v,d, flag = 1)=
{
  local(A);
  if(flag,
    A=concat(matrix(length(v),d+1,j,k,numerator(v[j])*(j-1)^(k-1)),
    matrix(length(v),d+1,j,k,-denominator(v[j])*(j-1)^(k-1)));
    A=qflll(A,4)[1],
    A=concat(matrix(length(v),d+1,j,k,v[j]*(j-1)^(k-1)),
    matrix(length(v),d+1,j,k,-(j-1)^(k-1)));
    A=matker(A));
  if(A,
    A=A[,1];
    sum(n=0,d,A[n+d+2]*x^n)/sum(n=0,d,A[n+1]*x^n),
    0)
}
```

The input for this function is a vector v with the initial terms of the sequence to be tested and d is a bound on the degree of the polynomials we are looking for. The output is a rational function $r(x) = p(x)/q(x)$ with $\deg p, \deg q \leq d$ such that $v[n+1] = r(n)$ for $n = 0, 1, \ldots$ if it exists and 0 otherwise.

4.2.3 Constant term recursion

If a sequence a_n satisfies a linear recurrence with constant terms

$$c_d a_{n+d} + c_{d-1} a_{n+d-1} + \cdots + c_0 a_n = 0, \quad n \geq 0, \quad (4.49)$$

we could treat it as a special case of (4.7) and use seqlinrec. However, it is much better to apply the power series version of recognize using continued fractions (see §1.2.1). In fact, it is *exactly* the same routine except we replace sercontfrac for contfrac.

4 : Sequences

```
serrecognize(f) =
{
  local(m);
  m=contfracpnqn(sercontfrac(f));
  m[1,1]/m[2,1]
}
```

We can apply it, in particular, to eventually periodic sequences (their generating function is rational with a cyclotomic denominator), see [64]. To illustrate this we consider the eventually periodic sequence formed by the partial quotients in the continued fraction expansion of $\sqrt{D}$ with $D > 0$ a squarefree integer.

```
contfracrq(D) =
{
  local(v);
  v = contfrac(sqrt(D));
  serrecognize(Ser(v) + O(x^(length(v)-1)) )
}
```

Because of the way GP normalizes continued fractions we omit the last partial quotient as it might be off; see the warning in §1.2.1. For example,

```
? \p 9
   realprecision = 9 significant digits

? contfrac(sqrt(3))

   [1, 1, 2, 1, 2, 1, 2, 1, 2, 1, 2, 1, 2, 1, 2, 1, 3]
? \p 19
   realprecision = 19 significant digits

? contfrac(sqrt(3))

   [1, 1, 2, 1, 2, 1, 2, 1, 2, 1, 2, 1, 2, 1, 2, 1, 2,
   1, 2, 1, 2, 1, 2, 1, 2, 1, 2, 1, 3]
```

Here are the first few cases.

```
? for(d=2,25,if(issquarefree(d),print(d" ",contfracrq(d))))

2 (x + 1)/(-x + 1)
3 (x^2 + x + 1)/(-x^2 + 1)
5 (2*x + 2)/(-x + 1)
6 (2*x^2 + 2*x + 2)/(-x^2 + 1)
7 (2*x^4 + x^3 + x^2 + x + 2)/(-x^4 + 1)
10 (3*x + 3)/(-x + 1)
```

```
11  (3*x^2 + 3*x + 3)/(-x^2 + 1)
13  (3*x^5 + x^4 + x^3 + x^2 + x + 3)/(-x^5 + 1)
14  (3*x^4 + x^3 + 2*x^2 + x + 3)/(-x^4 + 1)
15  (3*x^2 + x + 3)/(-x^2 + 1)
17  (4*x + 4)/(-x + 1)
19  (4*x^6 + 2*x^5 + x^4 + 3*x^3 + x^2 + 2*x + 4)/(-x^6 + 1)
21  (4*x^6 + x^5 + x^4 + 2*x^3 + x^2 + x + 4)/(-x^6 + 1)
22  (-4*x^6 - x^5 - 2*x^4 - 4*x^3 - 2*x^2 - x - 4)/(x^6 - 1)
23  (4*x^4 + x^3 + 3*x^2 + x + 4)/(-x^4 + 1)
```

(Can you explain the symmetry of the coefficients in the numerator?)

Remark Computing the continued fraction of $\sqrt{D}$ with `contfrac(sqrt(D))` as we did is a poor idea for large D (why?). A better way is to use a specific algorithm, see for example [52], vol. 2, §4.5.3, Ex. 12.

4.2.4 A simple example

Consider the algebraic equation

$$f(u, x) = u^2(1 - u)^3 - x^2 = 0. \tag{4.50}$$

By Hensel's lemma (see §4.1.4) we see that there is a unique solution $U(x)$ in power series of the form $U(x) = x + \cdots$. We may calculate the first few terms of the expansion of this solution using the following modification of our implementation `hensel` (to account for the fact that we actually have $\partial f/\partial u(0, w_0) = 0$ in this case)

```
hensel1(f,w,n) =
{
  local(g,d,c,r);
  d=poldegree(w);
  c=subst(subst(deriv(f,u),x,0),u,w+O(x^(d+1)));
  r=valuation(c,x);
  c=-1/polcoeff(c,r);
    for(k=d+1,d+n,
       g=subst(f,u,w+O(x^(k+1)));
       g=polcoeff(g,k+r);
       w+=c*g*x^k);
  w+O(x^(d+n+1))
}
```

4 : Sequences

We find for example

```
? U=hensell(u^2*(1-u)^3-x^2,x,20)

x + 3/2*x^2 + 33/8*x^3 + 14*x^4 + 6783/128*x^5 + 429/2*x^6
+ 930465/1024*x^7 + 3978*x^8 + 585062907/32768*x^9 +
81719*x^10 + 99600496359/262144*x^11 + 1789515*x^12 +
35736206484587/4194304*x^13 + 40940460*x^14 +
6653319491239617/33554432*x^15 + 966955410*x^16 +
10187554953759811155/2147483648*x^17 + 23398424931*x^18 +
1992107582500227802155/17179869184*x^19 + 577092394824*x^20
+ 792504329751402794768385/274877906944*x^21 + O(x^22)

? U^2*(1-U)^3-x^2

O(x^23)
```

Can we identify sequence of coefficients of $U = \sum_{n \geq 0} U_n x^n$? As we mentioned in §4.1.4, U satisfies a linear differential equation with polynomial coefficients. Hence the U_n's satisfy a linear recurrence with polynomial coefficients (by [25] we know their degree is at most 4). Let us test this; to be safe, we take $d = r = 5$.

```
? seqlinrec(Vec(U),5,5)

[-3125/108*x^4 - 3125/18*x^3 - 10000/27*x^2 - 5875/18*x -
385/4, 0, x^4 + 9*x^3 + 269/9*x^2 + 391/9*x + 70/3, 0, 0,
0]
```

We conclude that most likely (can you prove it? see Ex. 17)

$$\frac{a_{n+2}}{a_n} = r(n), \quad r(x) = \frac{5}{12} \frac{(5x-2)(5x+2)(5x+4)(5x+6)}{(x+1)(x+2)(3x+2)(3x+4)} \quad (4.51)$$

Hence

$$U(x) = U_0(x^2) + xU_1(x^2) \quad (4.52)$$

with U_0, U_1 *hypergeometric* i.e., the ratio of two consecutive coefficients is a rational function of n. In retrospect, the coefficients U_n do seem to naturally divide according to the parity of n; look back at the first few terms given above and consider their denominators for example.

We could also determine $r(x)$ by applying our rational recognition routine seqisratnl to the ratios a_{n+2}/a_n.

```
? U= hensel(u^2*(1-u)^3-x^2,x,40);v=Vec(U);
? v=vector(20,n,v[n+2]/v[n]);
? seqisratnl(v,4)
```

```
(3125*x^4 + 18750*x^3 + 40000*x^2 + 35250*x + 10395)/
(108*x^4 + 972*x^3 + 3228*x^2 + 4692*x + 2520)
```

Note that though the following equations in polynomials p and q

$$q(n)a_{n+2} - p(n)a_n = 0, \quad \frac{a_{n+2}}{a_n} = \frac{p(n)}{q(n)}, \quad n = 1, 2, \ldots \tag{4.53}$$

are formally equivalent (and we can solve the first using `seqlinrec` and the second with `seqisratnl`), in practice it might better to use the second equation as the numbers involved would typically have smaller height. Finally, notice that the denominators of U_n are only powers of 2 (compare with Ex. 6.19).

Remark For a series of much more sophisticated algorithms to deal with the recognition of sequences and more see [73].

4.3 Exercises

1. Try `trinomial2(0)` and explain the problem.
2. Prove (4.38) and (4.39).
3. Investigate the denominators of the coefficients c_k and prove what you find.
4. Yet another way to compute the sequence of trinomial numbers is by means of Lagrange's inversion formula [52], vol. 2, Ex. 8, p. 533. If

$$z = t/(1 + t + t^2) = t - t^2 + t^4 - t^5 + \cdots$$

then

$$\frac{z}{t\,dz/dt} = 1 + t_1 z + t_2 z^2 + \cdots$$

Check this identity and write a function that computes $t_0, t_1, \ldots, t_n$ using it.

5. Find numerically what degree of approximation to $\sum_{n \geq 0} t_n x^{n+1}$ we get from

   ```
   L=-2+2/x;contfracper([0,L/2],[-L/4,L],n)
   ```

   as a function of n and prove it.
6. Compute the continued fraction expansion of $\sqrt{x^{-4} + 10x^{-2} - 96x^{-1} - 71}$ and verify it is (eventually) periodic with period 14.
7. Compute the algebraic equation satisfied by the series U_0 and U_1 of §4.2.4.
8. Using `seralgdep` make a guess for the algebraic equations satisfied by the following power series and then prove that your guess is correct.

$$\sum_{n=0}^{\infty} \binom{2n}{n} x^n, \quad \sum_{n=0}^{\infty} \frac{1}{n+1} \binom{2n}{n} x^n,$$

(the coefficients in the second series are the famous *Catalan numbers* Sloane A108).

9. Repeat the previous exercise for

$$u_r := 1 + \sum_{n=1}^{\infty} \frac{(rn)!}{((r-1)n+1)!\,n!} x^n, \quad r = 1, 2, 3 \ldots \tag{4.54}$$

and describe (and eventually prove) the general form of the algebraic equation satisfied by u_r.

10. Use the previous exercise and Hensel's lemma to prove that
$$C_{r,n} := \frac{1}{(r-1)n+1}\binom{rn}{n} \in \mathbb{Z}$$
for all r and n. Prove, furthermore, that for r a prime number $C_{r,n} \equiv 1 \bmod r$ if n is a power of r and $C_{r,n} \equiv 0 \bmod r$ otherwise.

11. Find and prove a simple formula for the coefficients in the Taylor expansion of $1/u_r$ with u_r given by (4.54).

12. Find the algebraic equation satisfied by
$$1 + \sum_{n=1}^{\infty}\binom{rn}{n}x^n, \qquad r = 1, 2, 3 \ldots$$

13. Let $\sum_{n\geq 1} a_n x^n$ be the inverse power series (in the sense of composition) of xe^{-x}. Find and prove a formula for the coefficients a_n. (You may use `serreverse(x*exp(-x))` in GP to compute the first few a_n.)

14. Delannoy numbers, Sloane A1850, describe the number of paths from the southwest corner of a rectangular grid to the northeast corner, using only single steps north, northeast, or east. The sequence begins
$$a_n = 1, 3, 13, 63, 321, 1683, \ldots$$
and has the closed formula
$$a_n = \sum_{k=0}^{n} \binom{n}{k}\binom{n+k}{k}.$$

(a) Find the algebraic and differential equation satisfied by the generating function $\sum_{n=0}^{\infty} a_n x^n$.
(b) Write a GP script that computes a_n using the linear recurrence deduced from part (a).
(c) Find an asymptotic expansion for a_n.

15. Repeat the previous problem for the sequence a_n of Motzkin numbers, Sloane A001006, beginning with $1, 1, 2, 4, 9, 21, 51, \ldots$ One combinatorial description (of many) of a_n is the number of sequences of non-negative integers $(s_0, s_1, \ldots, s_n)$ with $s_0 = s_n = 0$ and $|s_i - s_{i-1}| \leq 1$ for $i = 1, 2, \ldots, n$. The Motzkin numbers have the closed form
$$a_n = -\frac{1}{2}\sum_{k=0}^{n+2}(-3)^k\binom{\frac{1}{2}}{k}\binom{\frac{1}{2}}{n+2-k}.$$

16. Write a GP script that will list for a given $n \in \mathbb{N}$ all the sequences $(s_0, s_1, \ldots, s_n)$ in the combinatorial description of Motzkin numbers given in the previous problem.

17. Consider the power series solution $U(x) = x + \cdots$ to the equation
$$u^r(1-u)^s - x^r = 0$$
for non-negative integers r and s. Investigate the relation of $U(x)$ to hypergeometric series, as we did in §4.2.4 for $r = 2$ and $s = 3$, for other small values of r and s. Prove the facts that you discover.

4.3 Exercises

18. Find the linear recurrence satisfied by the Apéry numbers, Sloane A005259,

$$A(n) := \sum_{k=0}^{n} \binom{n+k}{k}^2 \binom{n}{k}^2 = 1, 5, 73, 1445, 33001, 819005, \ldots$$

19. *Euler's Exemplum Memorabile Inductionis Fallacis* Let F_n be the Fibonacci sequence $F_{-1} = 1, F_0 = 0, F_1 = 1, \ldots$

$$1, 0, 1, 1, 2, 3, 5, 8, \ldots, \qquad F_{n+1} = F_n + F_{n-1}.$$

Euler noticed that

$$3t_{n+1} - t_{n+2} = F_n(F_n + 1), \qquad n = 0, \ldots, 8 \qquad (4.55)$$

(with t_n the trinomial number §4.1) but fails for $n = 9$. Check this out. How do you explain the coincidence (4.55)?

20. Check the following statements numerically.
    (1) Let a_n be the sequence defined by

$$a_{n+2} := \left\lfloor \frac{a_{n+1}^2}{a_n} + 1 \right\rfloor, \quad (n \geq 0), \quad a_0 = 8, a_1 = 55$$

where $\lfloor x \rfloor$ is the integer part of x. Let b_n be the n-th coefficient in the Taylor expansion about the origin of the rational function

$$\frac{8 + 7x - 7x^2 - 7x^3}{1 - 6x - 7x^2 + 5x^3 + 6x^4} = 8 + 55x + 379x^2 + 2612x^3 + \cdots$$

Then

$$a_n = b_n, \qquad n = 0, 1, \ldots, 11055$$

but fails for $n = 11056$.
    (2) Let

$$a_n := \left\lfloor \frac{a_{n+1}^2}{a_n} + \frac{1}{2} \right\rfloor, \quad (n \geq 0), \quad a_0 = 3, a_1 = 10$$

then

$$1 + \sum_{n \geq 0} a_n x^{n+1} = \frac{1}{1 - x(3 + x)} = 1 + 3x + 10x^2 + 33x^3 + 109x^4 + \cdots$$

21. Consider the 6-Somos, Sloane A006722,

$$1, 1, 1, 1, 2, 3, 7, 23, 59, 314, \cdots,$$

given by the following GP script.

```
somos6(n) =
{
    local(a);
    a=vector(n+1);
    for(k=1,6,a[k]=1);
    for(k=6,n,
```

```
        a[k+1] = (a[k]*a[k-4]+a[k-1]*a[k-3]+a[k-2]^2)/a[k-5]);
        a
}
```

Check with various n's to convince yourself that the sequence obtained, despite what its definition would lead one to believe, appears to give all integers.

22. (Hard) The Göbel sequence, Sloane A003504, is defined recursively as

$$a_0 := 1, \quad a_n := \frac{1}{n}(1 + a_0^2 + \cdots + a_{n-1}^2)$$

and starts $1, 2, 3, 5, 10, 28, 154, \ldots$. Try to find as many terms of this sequence as you can. The a_n appear to be all integers but this is not true.

23. Consider the following random Fibonacci sequence

$$x_{n+1} := x_n \pm \beta x_{n-1}, \quad x_0 = x_1 = 1,$$

where $\beta \in \mathbf{R}$ is a fixed parameter and ± 1 is chosen randomly with equal probability. Study numerically the behavior of $\lambda_n := \frac{1}{n} \log |x_n|$ as a function of β.

24. Identify the following sequence

$$a_n := 2^{(n-1)^2} \prod_{r=1}^{n} \frac{r^r (2r-1)^{2(r-1)}}{(n+r-1)^{n+r-2}}, \quad n \text{ odd}.$$

25. Let $M(x) := 1/(2 \lfloor x \rfloor + 1 - x)$ and define x_n recursively by $x_0 = 1$ and $x_{n+1} = M(x_n)$. What can you say about this sequence?

26. Find a formula for the inverse of the function M of the previous exercise.

27. Let $S(x) := (1+x)x - \lfloor x/2 \rfloor (2x+1)$. Study the sequence $x_{n+1} = S(x_n)$ for various starting points $x_0 \in \mathbf{Z}$.

5 Combinatorics

In this chapter we will discuss a basic programming idea and illustrate it with several examples.

Given positive integers $0 < m_1 < m_2 < \cdots < m_N$ and a positive integer m consider the problem of finding all non-negative integers $c_1, \ldots, c_N$ such that

$$c_1 m_1 + \cdots + c_N m_N = m. \tag{5.1}$$

We want a reasonable algorithm to solve this equation.

This kind of equation appears frequently. For example, if $m_j = j$ for $j = 1, 2, \ldots, N$ and $m = N$ then the solutions correspond to *partitions* of N; i.e., to sequences $\lambda_1 \geq \lambda_2 \geq \cdots$ of positive integers such that $N = \lambda_1 + \lambda_2 + \cdots$. Indeed, we can associate to a solution $(c_1, c_2, \ldots, c_N)$ the partition with c_j λ's equal to j for each j with $c_j > 0$ and vice-versa. The study of partitions is a very broad subject with connections to many areas of mathematics (e.g. representation theory of the symmetric groups, etc.); we will consider them in a bit more detail in the next section.

In general, we may think of a solution of (5.1) as a generalized partition of m, where m gets decomposed in c_j parts of size m_j. Concretely, we may think (5.1) as all ways of making a total of m out of coins with denominations $m_1, \ldots, m_N$.

Another problem that we will consider is the following one: find all polynomials in $\mathbb{Z}[x]$ of a given degree n whose roots consists only of roots of unity. As a sub-problem of this we will also solve the problem: find all $\nu \in \mathbb{N}$ such that $\phi(\nu) \leq n$, where ϕ is Euler's ϕ function.

5.1 Description of the basic algorithm

The best way to describe the main algorithmic idea of this chapter is by means of the concept of a *stack*. A stack is a storage structure in which: we may either place a new object at the top of the stack or we may remove the object currently at the top of the stack.

Actually, what we will use is not, strictly speaking, a stack. We will allow ourselves to take a peek at the top object without popping it and immediately pushing it back.

In our algorithm to solve (5.1) given below the stack will be represented by the two vectors `sm` and `sj` with the top of the stack determined by the current value of the index `k`. The objects in the stack for $1 < i \leq k$ are pairs of integers `sm[i]`, `sj[i]`; the first is the value of $s = c_1 m_1 + \cdots c_N m_N$ at the i-th level in the stack and the second is the index $j = 1, \ldots, N$ such that `sm[i] = sm[i-1] + mv[j]` (in other words, the index j of the new m_j we are adding on to go from the $(i-1)$-th value of s to the i-th value). The first object (at the bottom of the stack) is special and is set to `sm[1]=0, sj[1]=0`.

We start with the stack consisting of the first object only (so `k=1`). From a given stage of the process we proceed by constructing a new object `s, j` and testing whether `s <= m`. If this holds we have a linear combination $s \leq m$ of the m_j's, hence an object we want and put in the (top of the) stack; if it does not hold it means that we went too far. Since we assume that the m_j's are given in ascending order we know that further values of m_j down the list will be too big as well. Hence, we backtrack and remove objects from the top of the stack that we do not need anymore (by doing `k-`) until we find an object `sm[i], sj[i]` with `sj[i] < N` (i.e. such that there still is some candidate m_j to be tested at that point). The process ends when the stack is empty `k=0`.

Note that in fact we will be generating all solutions to the linear inequality

$$c_1 m_1 + \cdots c_N m_N \leq m; \qquad (5.2)$$

we simply keep those which actually give an equality.

For example, in the case of (5.3) the sequence of solutions to

$$c_1 2 + c_2 3 + c_3 5 \leq 10$$

would be generated in the following order (here given in the

$$(m_1, m_1, \ldots, m_2, m_2, \ldots)$$

format):

() (2) (2,2) (2,2,2) (2,2,2,2) (2,2,2,2,2) (2,2,2,3) (2,2,3)

(2,2,3,3) (2,2,5) (2,3) (2,3,3) (2,3,5) (2,5) (3) (3,3) (3,3,3)

(3,5) (5) (5,5)

Remark Note that though we assumed that the m's are increasing the algorithm will work just as well if we only assume that they are non-decreasing, i.e., $0 < m_1 \leq m_2 \leq \cdots \leq m_N$. We will see in §5.4 how this can be useful.

Here is a GP version of this algorithm.

```
lineq(mv,m,flag) =
{
  local(k,j,sm,sj,s,sv, S = [], N = length(mv));
  k = j = 1;
  sm = sj = vector(m\mv[1]+1);

  while(k, /* We are done when stack is empty */

  /* Create new object s,j */
    s = sm[k]+mv[j];

    if (s > m, /* Object too large, backtrack */
      until(j <= N, j = sj[k]+1; k--); /* k-- = pop last */
      next);

  /* Object suitable, push it at the top of the stack */
    k++; sm[k]=s; sj[k]=j; /* here s <= m, j <= N */

    if (s < m, next); /* Not a solution to linear equation,
    go on */

  /* Actual solution, add to solution vector */
    if (flag,
      sv = vector(k-1,l, mv[sj[k-l+1]])
    , /* else */
      sv = vector(N);
      for(l=2,k, sv[sj[l]]++));

    S = concat(S,[sv]));
  S
}
```

The algorithm takes as an input a vector mv of positive integers in ascending order and a target integer m; it outputs a vector with the solutions in one of two forms. If the optional input flag is zero (the default) each solution is given as the vector $(c_1, \ldots, c_N)$; otherwise it is given as a vector whose every entry is one of the mv[j] 's (in descending order) with total sum m. It is convenient to have both ways of representing the solutions.

Remark In writing and testing this function we replaced the line

```
until(j <= N, j = sj[k]+1; k--); /* k-- = pop last */
```

by

```
until(j <= N, /* pop last */
   j = sj[k]+1;
   sm[k]=0; sj[k]=0; k--);
```

The purpose was to clear the vectors representing the stack. This extra line `sm[k]=0; sj[k]=0` is technically unnecessary since the values we are clearing are, if anything, going to be replaced by newer objects. However, to test the code if we modify it for example, it proved useful.

Note also that lineq fails if $m = 0$ (do you see why?)

For example, all solutions in non-negative integers to the equation

$$c_1 2 + c_2 3 + c_3 5 = 10 \tag{5.3}$$

are

```
? lineq([2,3,5],10)

[[5, 0, 0], [2, 2, 0], [1, 1, 1], [0, 0, 2]]
```

or in the other format

```
? lineq([2,3,5],10,1)

[[2, 2, 2, 2, 2], [3, 3, 2, 2], [5, 3, 2], [5, 5]]
```

As a check we can try the following

```
? N=50; sum(n=1,N-1, x^n * length(lineq([2,3,5],n))),
1+O(x^N))*(1-x^2)*(1-x^3)*(1-x^5)

1 + O(x^50)
```

(Why is this a check?)

As another example, all partitions of $m = 5$ are obtained as follows

```
? lineq([1,2,3,4,5],5,1)

[[1, 1, 1, 1, 1], [2, 1, 1, 1], [3, 1, 1], [2, 2, 1],
[4, 1], [3,2], [5]]
```

Remark It is known that for a fixed set of integers $m_1,\ldots,m_N$ and for every sufficiently large integer m there is a solution to (5.1). The problem of finding the largest number m_0 for which there is no solution is known as the *money changing problem*. There is no known closed formula for m_0 if $N > 3$.

5.2 Partitions

It is useful to have an algorithm specifically tailored made for the case of partitions. Modifying the previous one for example we get the following.

```
part(m) =
{
  local(k,j,sm,sj,s, S = []);
  k = j = 1;
  sm = sj = vector(m+1);

  while(k,
        s = sm[k]+j;
        if (s > m,
        until(j <= m, j = sj[k]+1; k--);
        next);

        k++; sm[k]=s; sj[k]=j;

        if (s == m,
        S = concat(S, [vector(k-1,1, sj[k-1+1])])));
  S
}
```

For example

```
? part(5)

[[1, 1, 1, 1, 1], [2, 1, 1, 1], [3, 1, 1], [2, 2, 1],
[4, 1], [3, 2], [5]]
```

We can easily modify the routine `part` so that it returns all partitions of numbers $m \leq n$. It is enough to initialize the solution vector as S=vector(n,1,[]) and change the two lines

```
        if(s == m,
        S=concat(S, [vector(k-1,1,sj[k-1+1])])),
```

to

```
S[s]=concat(S[s],[vector(k-1,1,sj[k-1+1])]),
```

As an alternative we could also use the following implementation (from the FAQ of PARI-GP)

```
part2(n) = for(i=1, n, AUX(n-i, i, []))

AUX(n, m, v) =
{
  v = concat(v,m);
  if(n, for(i=1, min(m,n), AUX(n-i, i, v))
    , print(v));
}
```

where the stack is replaced by a recursion.

5.2.1 The number of partitions

The number of partitions of n is denoted by $p(n)$. This number grows quite fast with n; it is known that as n goes to infinity

$$p(n) \sim \frac{1}{4n\sqrt{3}} e^{\pi\sqrt{2n/3}}. \qquad (5.4)$$

There is, in fact, an explicit convergent series, due to Rademacher (see [77]), with the right-hand side of (5.4) as its first term (see below).

As we mentioned above, the algorithm actually finds the partitions of all $m \leq n$; hence we can easily modify part to give $p(m)$ for all $m \leq n$

```
partnum(m) =
{
  local(k,j,sm,sj,s, S = vector(m));
  k = j = 1;
  sm = sj = vector(m+1);

  while(k,
      s = sm[k]+j;
      if (s > m,
        until(j <= m, j = sj[k]+1; k--);
      next);

      k++; sm[k]=s; sj[k]=j; S[s]++);

  S
}
```

5.2 Partitions

For example,

```
? partnum(20)
```

[1, 2, 3, 5, 7, 11, 15, 22, 30, 42, 56, 77, 101, 135, 176, 231, 297, 385, 490, 627]

There is however a much better way of doing this. As noted by Euler,

$$\sum_{n=0}^{\infty} p(n)q^n = \frac{1}{\prod_{n=1}^{\infty}(1-q^n)}, \qquad (5.5)$$

(with the agreement that $p(0) := 1$), which gives a quick way of computing $p(m)$ for all $m \leq n$ as follows

```
partnum1(n) = Vec(1/prod(m=1,n,1-x^m+O(x^(n+1)))-1)
```

In fact, there is an even a better way of doing this. Euler proved the following remarkable expansion of the infinite product in (5.5) (see [46])

$$\prod_{n=1}^{\infty}(1-q^n) = \sum_{n \in \mathbb{Z}}(-1)^n q^{\frac{1}{2}n(3n+1)}. \qquad (5.6)$$

The right-hand side is a kind of *theta series* as in §1.4.5 and is basically Dedekind's eta function, see §1.2.3. The exponents of q in the series $0, 1, 2, 5, 7, \ldots$ are called *pentagonal numbers* Sloane A001318.

From a computational point of view the right-hand side of (5.6) is very convenient since roughly a constant times $\sqrt{n}$ terms give an accuracy of $O(q^n)$. Here is an implementation of Euler's formula in GP (we use the variable x instead of q). Note the convoluted way in which the terms $(-1)^n x^{\frac{1}{2}n(3n+1)}$ are generated; we leave it as an exercise for the reader to check it. The GP has a built-in function eta(x) also computes the infinite product of (5.6) as a power series.

```
etaps(n) =
{
  local(z,t1,t2,q1,q2,q3);
  z=1+O(x^(n+1));t1=1;t2=1;
  q1=x; q2=x^2; q3=x^3;
  until(poldegree(t2)>n,
    t1*=-q1;t2*=-q2;
    z+=t1+t2;
    q1*=q3;q2*=q3);
  z
}
```

With this function we can now write

```
partnum2(n) = Vec(1/etaps(n)-1)
```

or simply, using GP's built-in function,

```
partnum3(n) = Vec(1/eta(x + O(x^(n+1))) - 1)
```

Finally, here is an implementation of Rademacher's formula for $p(n)$ (written by R. Stephan), which is very good if we want and individual value of $p(n)$.

```
partnum4(n) = round(sum(q=1,5 + 0.24*sqrt(n),L(n,q)*Psi(n,q)))

Psi(n, q) =
{
  local(a, b, c);
  a=sqrt(2/3)*Pi/q; b=n-1/24; c=sqrt(b);
  (sqrt(q)/(2*sqrt(2)*b*Pi))*(a*cosh(a*c)-(sinh(a*c)/c))
}

G(h, q) = if(q<3,0,sum(k=1,q-1,k*(frac(h*k/q)-1/2)))

L(n, q) =
{
  if(q==1,1,
    sum(h=1,q-1,
      if(gcd(h,q)>1,0,cos((G(h,q)-2*h*n)*Pi/q))))
}
```

5.2.2 Dual partition

A partition $\lambda = (\lambda_1, \lambda_2, \ldots)$ has an associated *dual partition* $\lambda' = (\lambda'_1, \lambda'_2, \ldots)$. The easiest way to describe this is with pictures. We represent a partition by a Young diagram formed of λ_j boxes on the j-th row flushed left. The dual partition then corresponds to reflecting the diagram along a

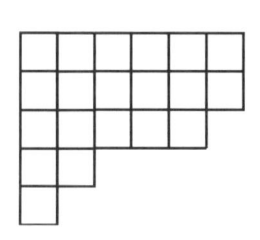

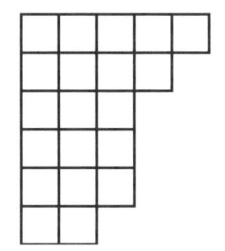

Figure 5.1 *A partition and its dual*

NW-SE diagonal. For example (see Fig. 5.1), the partition (6, 6, 5, 2, 1) has dual partition (5, 4, 3, 3, 3, 2).

We leave it as an exercise to verify that the following routines compute the dual of a partition.

```
partdual(v)=
{
    local(w);
    w=vector(v[1]);
    for(j=1,length(v),
    for(i=1,v[j], w[i]++));
    w
}

partdual1(v) =
{
  local(last, w = vector(v[1]));
    w[1] = last = length(v);
      for (i=2, v[1],
        while (v[last] < i, last--);
        w[i] = last);
      w;
}
```

For example,

```
? partdual([3,2,2,1])

 [4, 3, 1]
```

5.3   Irreducible representations of S_n

In this section we will consider various questions associated to irreducible representations of the group S_n of permutations of n objects. In particular, we will write a GP routine to compute the character table of S_n. We should warn the reader that the number of computations in the routines in this section grows exponentially with n making them unfeasible fairly quickly. Nevertheless, in practice, they can still be useful both in themselves and as examples of programming solutions to the problems they set out to solve. For a bigger scale calculations with groups and characters one should use more specific and powerful tools (for example as in GAP [38]). For more details see [37].

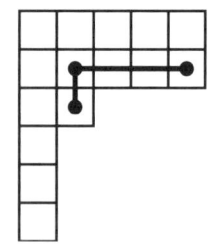

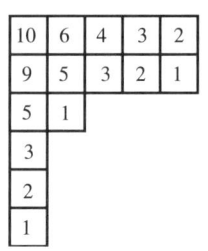

Figure 5.2 *Hook length*

5.3.1 Hook formula

The partitions of a given positive integer n parameterize the irreducible representations of the symmetric group S_n. The dimension f_λ of the representation associated to λ has the following beautiful closed formula [36], [37], §4.1, p. 50.

$$f_\lambda = \frac{n!}{\prod_x h(x)} \qquad (5.7)$$

where x runs through the boxes of the Young diagram associated to λ and $h(x)$ is the *hook length* of x, defined as the number of boxes in the diagram to the right and below x (including itself). For example, each box in the partition $(5, 5, 2, 1, 1, 1)$ has hook length as indicated in the right of Fig. 5.2 (on the left we show the hook of the $(2, 2)$ box, which has length 5).

As a further exercise we leave to the reader to verify that the following algorithm computes f_λ for a given partition.

```
partdim(v) =
{
  local(w,ct,dim);
  w=partdual(v);
  dim=1;ct=1;
    for(j=1,v[1],
      for(k=1,w[j],
        dim=dim*ct/(v[k]+w[j]-j-k+1);
        ct++));
  dim
}
```

For example, for $\lambda = (4, 3, 1)$ a partition of 8 we find $f_\lambda = 8!/6 \cdot 4 \cdot 4 \cdot 2 \cdot 3 = 70$ and indeed

```
? partdim([4,3,1])

70
```

For all partitions of $n = 5$ say we find the following dimensions.

```
? pv=part(5)

[[1, 1, 1, 1, 1], [2, 1, 1, 1], [3, 1, 1], [2, 2, 1],
 [4, 1], [3, 2], [5]]

? vector(length(pv),k,partdim(pv[k]))

[1, 4, 6, 5, 4, 5, 1]
```

As a substantial check in our routines we can verify that

$$n! = \sum_\lambda f_\lambda^2, \tag{5.8}$$

where the sum is over all partitions of n, for small values of n (this formula is the special case for the group S_n of a general formula relating the size of a group and the dimension of its irreducible representations).

```
? for(n=2,10,pv=part(n);print1(sum(k=1,length(pv),
partdim(pv[k])^2)/n!,""))

1  1  1  1  1  1  1  1  1
```

5.3.2 The Murnaghan–Nakayama rule

This rule (see [37], §4.3, p.59) gives a recursive way of computing the values of the irreducible characters of S_n. To each partition λ of n there is an associated irreducible representation with character χ^λ. On the other hand, the conjugacy classes of S_n are also parameterized by the partitions of n; if $\mu = (\mu_1, \mu_2, \ldots)$ is such a partition the corresponding conjugacy class is that of $\sigma_1 \sigma_2 \cdots \in S_n$ with σ_i disjoint cycles of length μ_i. We denote by χ^λ_μ the value of χ^λ on the conjugacy class indexed by μ.

In order to describe the recursion we need to introduce the concept of a *rim hook* of a partition λ of n; it consists of a continuous portion of the boundary of λ, which after its removal yields another partition. We call the number of boxes in the rim hook its *length*.

For example, the partition $(6, 5, 5, 4, 2, 2, 1)$ has as one possible rim hook of length 5 the shaded boxes (see Fig. 5.3); removing it leaves the partition $(6, 4, 3, 2, 2, 2, 1)$

Given another partition μ of n we consider chains of partitions

$$C : \lambda^{(0)} = \lambda, \lambda^{(1)}, \ldots, \lambda^{(k)} = 0, \tag{5.9}$$

134   **5 : Combinatorics**

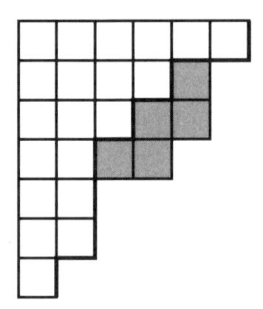

Figure 5.3 *A rim hook*

where $\lambda^{(i)}$ is obtained from $\lambda^{(i-1)}$ by removing a rim hook of length μ_i. To such a chain we associate a sign sgn(C) defined as the product of $(-1)^{r-1}$ over all rim hooks of C, where r is the number of rows in each rim hook. We then have

$$\chi_\mu^\lambda = \sum_C \text{sgn}(C) \qquad (5.10)$$

where C runs over all chains.

To implement this formula in GP we first define a function (coded by T. Geldon) that given a partition λ and a number $m \in \mathbf{N}$ returns all partitions obtained from λ by removing a rim hook of length m. In fact, we also need to keep track of the sign needed to compute sgn(C). Hence the input of our function actually is a pair (λ, e) and an integer $m \in \mathbf{N}$, where λ is a partition and e is an integer modulo 2; the output consists of all pairs (λ', e') with λ' a partition obtained from λ by removing a rim hook of length m and $e' \equiv e + r - 1 \mod 2$, where r is the number of rows of the hook removed.

```
partrmrim(vr,m) =
{
  local(Solutionv,newv,v,neww,w,row,rimlen,col,r,currlen);
  v=vr[1];
  w=partdual(v);
  Solutionv=[];

  r=0; newv=v;

/* Col represents the column where the current rim begins
   Row represents the row where the current rim ends */

  col=1; row=w[col];
```

```
/* Start off having removed the last row */

  newv[row] = 0;
  rimlen = v[row];
  if(rimlen == m, Solutionv=concat(Solutionv,[[newv,vr[2]]]));

/* could say "if rim==m, Sol=this, Sol=[]" and take out
    init of Soln*/

  while(row > 1 || (rimlen >= m && v[row] != col),
    if ((rimlen < m) || (v[row] == col),

/* Move up a row to increase the length if the length is too
    small, or if moving over a column will delete the rim */

      row--;
      newv[row] = v[row+1]-1;
      rimlen = rimlen + v[row]-newv[row],

    /* else: move over a column to decrease the length */

      neww=partdual(newv);
        if (length(neww)<col, currlen = 0, currlen = neww[col]);

      rimlen = rimlen + currlen - w[col];
        if (currlen > 0, neww[col] = w[col],
         neww=concat(neww,w[col]));
      newv = partdual(neww);
      col++);
    if(rimlen == m,
      Solutionv=concat(Solutionv,[[newv,(vr[2]+w[col]-row)%2]])));
  Solutionv
}
```

For example there are only two rim hooks of length 4 for the partition $(3, 2, 2, 1)$ which, when removed, leave the partitions $(3, 1)$ and $(1, 1, 1, 1)$ respectively. Since both hooks have three rows each the exponent e does not change. Computing with our GP function we find that indeed

```
? partrmrim([[3,2,2,1],0],4)

[[[3, 1, 0, 0], 0], [[1, 1, 1, 1], 0]]
```

Note the harmless fact that the partitions in the output are given as vectors of the same length as the input without trimming any trailing zeros (in our example we have [3,1,0,0] and not [3,1]).

Similarly, for the partition (6, 5, 5, 4, 2, 2, 1) of Fig. 5.3 we find the rim hook of length 5 depicted and one other.

```
? partrmrim([[6,5,5,4,2,2,1],0],5)

  [[[6, 5, 5, 1, 1, 1, 1], 0], [[6, 4, 3, 2, 2, 2, 1], 0]]
```

With `partrmrim` as the basic recursive step we may now implement the Murnaghan–Nakayama rule (5.10)

```
partchar(v,u) =
{
  local(V,W);
  V=[[v,0]];
    for(i=1,length(u),
      W=[];
        for(j=1,length(V),
          W=concat(W,partrmrim(V[j],u[i])));
      V=W);

  sum(k=1,length(W),if(W[k][2],-1,1))
}
```

Finally, we can compute the whole character table of S_n with the following function.

```
partchartable(n) =
{
  local(w);
  w=part(n);
  matrix(length(w),length(w),j,k,partchar(w[j],w[k]))
}
```

For example, for $n = 5$ we find

```
? partchartable(5)

[1 -1 1 1 -1 -1 1]

[4 -2 1 0 0 1 -1]

[6 0 0 -2 0 0 1]

[5 -1 -1 1 1 -1 0]

[4 2 1 0 0 -1 -1]

[5 1 -1 1 -1 1 0]

[1 1 1 1 1 1 1]
```

(Here the ordering of the characters and conjugacy classes is that determined by those of the partitions of 5 as given by the function part.)

We should check our routines. We can check, for example, that the character table we compute satisfies the standard orthogonality relations. Namely,

$$\sum_{\mu} \frac{1}{z_\mu} \chi_\mu^\lambda \chi_\mu^{\lambda'} = \delta_{\lambda,\lambda'}, \tag{5.11}$$

where

$$z_\mu = 1^{m_1} m_1! 2^{m_2} m_2! \cdots \tag{5.12}$$

with m_i the multiplicity of $i \in \mathbf{N}$ in the partition μ. Here z_μ is such that the conjugacy class corresponding to μ has size $n!/z_\mu$. We may compute these quantities with the following GP routine (a direct implementation of (5.12))

```
partz(v)=
{
  local(w);
  w=vector(v[1]);
  for(j=1,length(v),
    w[v[j]]++);
  prod(j=1,length(w),
    j^w[j]*w[j]!)
}
```

Here is then a GP function to check the orthogonality relations.

```
partcharcheck(n)=
{
  local(w,v,M);
  w=part(n);
  v=vector(length(w),k,partz(w[k]));
  M=matrix(length(w),length(w),j,k,partchar(w[j],w[k]));
  matid(length(w))==M*matdiagonal(vector(length(v),k,1/v[k]))*M~
}
```

It returns 1 if (5.11) holds for the calculated values of χ_μ^λ and 0 if they do not. We find

```
? vector(8,n,partcharcheck(n))

[1, 1, 1, 1, 1, 1, 1, 1]
```

5.3.3 Counting solutions to equations in S_n

One may express the number of solutions to certain kinds of equations in a finite group G in terms of its irreducible characters. These ideas go back to Frobenius in the early development of the theory of representations of finite groups. We will only consider the following case [85]. Let $C_1, \ldots, C_m$ be conjugacy classes in G. Let

$$N(C_1, \ldots, C_m) := \#\{x_1 \cdots x_m = 1 \mid x_i \in C_i\}. \tag{5.13}$$

We then have the formula

$$N(C_1, \ldots, C_m) = \frac{|C_1| \cdots |C_m|}{|G|} \sum_\chi \frac{\chi(C_1) \cdots \chi(C_m)}{\chi(1)^{m-2}} \tag{5.14}$$

where the sum is over all irreducible characters of G and $\chi(C_i)$ denotes the value of χ on any element $x_i \in C_i$.

Using our routines for computing the character of S_n we can easily write a GP routine for computing $N(C_1, \ldots, C_m)$, for example, the following.

```
partnosol(ccv)=
{
  local(pv,m,n,nf);

  n=sum(k=1,length(ccv[1]),ccv[1][k]);
  pv=part(n);m=length(ccv);nf=n!;

  prod(k=1,m,nf/partz(ccv[k]))/nf*
  sum(j=1,length(pv),
    prod(k=1,m,
      partchar(pv[j],ccv[k]))/
      partdim(pv[j])^(m-2))
}
```

The input is a vector ccv of partitions of n representing the conjugacy classes $C_1, \ldots, C_m$ in S_n and the output is $N(C_1, \ldots, C_m)$.

As an illustration, let us compute all triples (C_1, C_2, C_3) of conjugacy classes in S_5 for which $x_1 x_2 x_3 = 1$ with $x_i \in C_i, i = 1, 2, 3$ has a solution in S_5. (Each line in the following list consists of the partitions of 5 giving the conjugacy classes C_1, C_2, C_3 and the corresponding number of solutions.)

```
? p5v=part(5);

? forvec(u=[[2,7],[1,7],[1,7]],w=vector(3,j,p5v[u[j]]);
N=partnosol(w);if(N,print(N,"\t",w)),1)
```

60	[[2, 1, 1, 1], [2, 1, 1, 1], [3, 1, 1]]
30	[[2, 1, 1, 1], [2, 1, 1, 1], [2, 2, 1]]
120	[[2, 1, 1, 1], [3, 1, 1], [4, 1]]
20	[[2, 1, 1, 1], [3, 1, 1], [3, 2]]
60	[[2, 1, 1, 1], [2, 2, 1], [4, 1]]
60	[[2, 1, 1, 1], [2, 2, 1], [3, 2]]
120	[[2, 1, 1, 1], [4, 1], [5]]
120	[[2, 1, 1, 1], [3, 2], [5]]
140	[[3, 1, 1], [3, 1, 1], [3, 1, 1]]
120	[[3, 1, 1], [3, 1, 1], [2, 2, 1]]
120	[[3, 1, 1], [3, 1, 1], [5]]
60	[[3, 1, 1], [2, 2, 1], [2, 2, 1]]
120	[[3, 1, 1], [2, 2, 1], [5]]
240	[[3, 1, 1], [4, 1], [4, 1]]
240	[[3, 1, 1], [4, 1], [3, 2]]
140	[[3, 1, 1], [3, 2], [3, 2]]
240	[[3, 1, 1], [5], [5]]
30	[[2, 2, 1], [2, 2, 1], [2, 2, 1]]
120	[[2, 2, 1], [2, 2, 1], [5]]
270	[[2, 2, 1], [4, 1], [4, 1]]
120	[[2, 2, 1], [4, 1], [3, 2]]
120	[[2, 2, 1], [3, 2], [3, 2]]
120	[[2, 2, 1], [5], [5]]
360	[[4, 1], [4, 1], [5]]
240	[[4, 1], [3, 2], [5]]
120	[[3, 2], [3, 2], [5]]
192	[[5], [5], [5]]

5.3.4 Counting homomorphism and subgroups

Consider the triangle group $T_{p,q,r}$ for $p, q, r \in \mathbf{N}$ generated by x, y, z with relations $x^p = 1, y^q = 1, z^r = 1$ and $xyz = 1$. It is known, for example, that $T_{2,3,5}$ is isomorphic to A_5. We can count the homomorphisms of $T_{p,q,r}$ into S_n by using (5.14). To simplify things a bit we will assume that p, q, r are distinct prime numbers. Then

$$\# \mathrm{Hom}(T_{p,q,r}, S_n) = 1 + \sum_{C_p, C_q, C_r} N(C_p, C_q, C_r), \tag{5.15}$$

where C_l, for the primes $l = p, q, r$, runs through the conjugacy classes of elements of order l in S_n. These classes have associated partition $(l, \ldots, l, 1, \ldots, 1)$ with $k = 1, 2, \ldots, [n/l]$ l's and $n - kl$ 1's.

Interchanging summations we see that

$$\# \mathrm{Hom}(T_{p,q,r}, S_n) = \frac{1}{n!} \sum_{\chi} \frac{s_p(\chi) s_q(\chi) s_r(\chi)}{\chi(1)}, \tag{5.16}$$

where χ runs over all irreducible characters of S_n and

$$s_l(\chi) := \sum_{C_l} \chi(C_l)|C_l|. \tag{5.17}$$

It follows from (5.12) that if C_l has associated partition $(l, \ldots, l, 1, \ldots, 1)$ as above then

$$|C_l| = \frac{n!}{l^k k!(n-lk)!}. \tag{5.18}$$

Here is the implementation of (5.16) in GP. The input is a three-component vector rv containing the primes p, q, r and the number n. The output is a polynomial in x of degree n whose k-th coefficient is $\# \mathrm{Hom}(T_{p,q,r}, S_k)$.

```
parttglhom(rv,n) =
{
/* Run a copy of part to get all partitions w of m <= n. */

  local(k,j,sm,sj,s,ct,w);
  k = j = 1;
  sm = sj - vector(n,1);

/* ct is the counting polynomial; set the intial value to
account for the trivial homomomorphism into S_k. */

  ct=sum(k=0,n,x^k)+O(x^(n+1));

    while(k,
      s = sm[k]+j;
      if (s > n,
        until(j <= n, j = sj[k]+1; k--);
        next);

      k++; sm[k]=s; sj[k]=j;

      w=vector(k-1,i,sj[k-i+1]);

/* With partition w compute the contribution to the total ct from
non-trivial conjugacy classes of orders rv[i] i=1,2,... */

        ct+=prod(i=1,length(rv),
           partpsum(rv[i],w))/partdim(w)*x^s/s!);
  ct
}
```

5.3 Irreducible representations of S_n

To simplify the programming the sum $s_l(\chi)$ is computed by the separate routine

```
partpsum(i,v) =
{
  local(s,k,n,z);

/* n varies: n, n-i, n-2i,... with |v| = n as initial value;
n!/z is the size of the corresponding conjugacy class C =
(i,...,i,1,...1) with k i's and (current value of) n 1's */

  n=sum(k=1,length(v),v[k]);
  k=0; z=1; s=0;

  while(i <= n,
    k++;
    z*=prod(j=1,i,n-j+1)/i/k;
    n-=i;
    s+=z*partchar(v,concat(vector(k,j,i),vector(n,j,1))));
  s
}
```

For example

```
? parttglhom([2,3,5],5)

1 + x + x^2 + x^3 + x^4 + 121*x^5 + O(x^6)
```

This answer makes sense: $T_{2,3,5} \simeq A_5$ is simple and hence has no non-trivial maps to S_k for $k < 5$; for $k = 5$ there are 120 embeddings $A_5 \hookrightarrow S_5$.

We may further combine this calculation with an instance of the *exponential formula* of combinatorics [89] (Ex. 5.13) valid for any finitely generated group Γ

$$1 + \sum_{n \geq 1} \# \operatorname{Hom}(\Gamma, S_n) \frac{x^n}{n!} = \exp\left(\sum_{n \geq 1} u_n(\Gamma) \frac{x^n}{n}\right), \quad (5.19)$$

where $u_n(\Gamma)$ denotes the number of subgroups of Γ of index n, to obtain the following routine to compute $u_n(T_{p,q,r})$. The input is of the same form and the output is $1 + \sum_{k=1}^n u_k(T_{p,q,r}) x^k$.

```
parttglsubgps(rv,n) =
{
/* Run a copy of part to get all partitions w of m <= n. */

  local(k,j,sm,sj,s,ct,w);
```

```
        k = j = 1;
        sm = sj = vector(n+1);

/* ct is the counting polynomial; set the intial value to
account for the trivial homomomorphism into S_k. */

        ct=sum(k=0,n,x^k/k!)+O(x^(n+1));

            while(k,
                s = sm[k]+j;
                if (s > n,
                    until(j <= n, j = sj[k]+1; k--);
                next);

            k++; sm[k]=s; sj[k]=j;

            w=vector(k-1,i,sj[k-i+1]);

/* With partition w compute the contribution to the total ct from
non-trivial conjugacy classes of orders rv[i] i=1,2,... */

            ct+=prod(i=1,length(rv),
                partpsum(rv[i],w))/partdim(w)*x^s/s!^2);

            x*deriv(ct,x)/ct
}
```

For example, for $T_{2,3,5}$ we find

```
? parttglsubgps([2,3,5],6)

x + 5*x^5 + 6*x^6 + O(x^7)
```

confirming that A_5 has no subgroups of index 2, 3, 4, five subgroups of index 5, and six of order 6. For $T_{2,3,7}$ we get

```
? parttglsubgps([2,3,7],12)

x + 14*x^7 + 8*x^8 + 9*x^9 + O(x^13)
```

The subgroups of index 7 and 8 are accounted for by the corresponding subgroups of the Klein group $\text{PSL}_2(\mathbf{F}_7)$ (see §2.5), which is a quotient of $T_{2,3,7}$.

5.4 Cyclotomic polynomials

As a final application of our algorithm `lineq` we consider listing all monic polynomials $P \in \mathbf{Z}[x]$ of a given degree $n \in \mathbf{N}$ all of whose roots are roots of unity. We will call such polynomials *cyclotomic polynomials*.

By a theorem of Kronecker a monic polynomial in $\mathbf{Z}[x]$ is cylotomic if and only if all of its roots are on the unit circle

$$P = \prod_{\nu \geq 1} \Phi_\nu^{\gamma_\nu}, \quad \gamma_\nu \in \mathbf{Z}_{\geq 0}, \tag{5.20}$$

where Φ_ν is the usual ν-th cyclotomic polynomial and only finitely many of the exponents γ_ν are non-zero.

Computing degrees we find

$$n = \sum_{\nu \geq 1} \gamma_\nu \, \phi(\nu), \quad \gamma_\nu \in \mathbf{Z}_{\geq 0}, \tag{5.21}$$

where ϕ is Euler's phi-function. Conversely, any solution to (5.21) corresponds to a unique cyclotomic polynomial of degree n.

We decompose our problem in two: first, we compute all ν such that $\phi(\nu) \leq n$; second, using a modification of `lineq` (that keeps track of the ν's with a given value of $\phi(\nu)$) we compute all solutions to (5.21) and hence find all cyclotomic polynomials of degree n. As we pointed out at the end of §5.1 the m's we input into `lineq` need not be strictly decreasing. This fact comes in handy here to solve (5.21) since there will, in general, be many ν's with the same value of $\phi(\nu)$.

5.4.1 Values of ϕ below a given bound

To compute all ν with $\phi(\nu) \leq n$ we can use the same ideas that we used for solving (5.2), now in a multiplicative form. Indeed, we know that

$$\phi(\nu) = \prod_j p_j^{r_j-1}(p_j - 1), \tag{5.22}$$

where ν has the prime decomposition $\nu = \prod_j p_j^{r_j}$. So we can generate the ν's by running over all primes p with $p - 1 \leq n$. We can easily modify the program `lineq` to do this; every time we move forward by using a prime p we either multiply the current value of ϕ by $p - 1$ if it is the first time we are using this particular prime or by p otherwise. We should also keep track of the value of v

$$v = \prod_j p^{r_j}.$$

Here is our version in GP.

```
eulerphilist(n) =
{
  local(p,pv,N,M,s,j,k,m,sph,sj,sm);
  pv=[];p=1;a=1;

  forprime(p=2,n+1,pv=concat(pv,p));

  M=ceil(log(n)/log(2))+2;
  S=vector(n\2,k,[]); N=length(pv);
  k=1;j=1;
  sph = vector(M,1,1);
  sm=sph; sj=vector(M);

  while(k,   /* We are done when stack is empty */

  /* Create new object s,m,j */
  s=sph[k]*(pv[j]-if(j==sj[k],0,1));
  m=sm[k]*pv[j];

    if(s > n,  /* Object too large, backtrack */
      until(j <= N,j = sj[k]+1; k--);   /* k-- = pop last */
    next);

    /* Object suitable, push it at the top of the stack */
    k++; sph[k] = s; sm[k] = m; sj[k] = j;  /* here s <= n,
                                                    j <= N */

    if(s > 1,S[s/2]=concat(S[s/2],m)));

  vector(length(S),k,vecsort(S[k]))
}
```

The objects in the stack now consist of triples s, m, j, kept in three vectors sph, sm, sj, corresponding respectively to $\phi(v), v, j$. Since the only possible values of $\phi(v)$ which are odd are $\phi(1) = \phi(2) = 1$ we give the output as a vector S whose r-th entry is the vector of values of v, in ascending order, such that $\phi(v) = 2r$.

5.4 Cyclotomic polynomials

For example,

```
? eulerphilist(8)

[[3, 4, 6], [5, 8, 10, 12], [7, 9, 14, 18], [15, 16, 20, 24, 30]]
```

so, in particular, all the $v \in \mathbf{N}$ with $\phi(v) = 8$ are $v = 15, 16, 20, 24, 30$.

As always when writing a program it is not a bad idea to test it in some reasonable cases in some other way, if we can. In the case at hand we can do the following random tests, for example.

```
? eulerphilist1(n,vv)=vv=[];for(m=1,10000,
      if(eulerphi(m)==n,vv=concat(vv,m)));vecsort(vv)
? eulerphicheck(n)=eulerphilist(2*n)[n]-eulerphilist1(2*n)

? for(k=1,10,n=random(100)+100;print(n,"   ",eulerphicheck(n)))

162  [0, 0, 0, 0, 0, 0, 0, 0]
121  []
190  [0, 0, 0]
132  [0, 0, 0, 0, 0, 0, 0, 0, 0, 0]
177  []
153  [0, 0]
189  [0, 0]
157  []
165  [0, 0]
121  []
```

However note that random checks might fail to reveal problems arising from special cases (for example, for small values of a parameter, say, where the generic behavior has not yet been attained).

5.4.2 Computing cyclotomic polynomials

We are now ready to compute all the cyclotomic polynomials of a given degree n. First we generate all pairs $(v, \phi(v))$ with $\phi(v) \leq n$ and then solve (5.21) and compute the corresponding cyclotomic polynomial.

```
polcyclolist(n) =
{
  local(nuv,nuv1,spol,sj,sdeg,N,j,k,pol,S,s);

    if(n > 1, nuv1 = eulerphilist(n));
  phv=[1,1];nuv=[1,2];
  S=[];
    for(k=1,length(nuv1),
      for(j=1,length(nuv1[k]),
```

```
              phv=concat(phv,2*k);
              nuv=concat(nuv,nuv1[k][j])));

   k=1; j=1; N=length(phv);

   spol = vector(N,k,1);
   sdeg = vector(N); sj = sdeg;

      while(k,
         /* Create new object s, pol, j */

         s = sdeg[k]+phv[j];
         pol = spol[k]*polcyclo(nuv[j]);

            if (s > n, /* Object too large, backtrack */
               until(j <= N, j = sj[k]+1; k--); /* k-- = pop last */
               next);

   /* Object suitable, push it at the top of the stack */
         k++; sdeg[k] = s; sj[k] = j; spol[k] = pol; /* here s <= n,
                                                         j <= N     */

            if (s < n, next); /* Not a solution to linear equation,
                                  go on */

      S=concat(S,pol));

   S
}
```

The vectors nuv and phv contain, respectively, the matching values of v and $\phi(v)$ ordered by ascending order of $\phi(v)$. They are computed after running eulerphilist(n), which, recall, gives the values of $(v,\phi(v))$ with $v > 2$ and $\phi(v) \leq n$. For example, for $n = 5$ we have

? eulerphilist(5)

[[3, 4, 6], [5, 8, 10, 12]]

from which polcyclolist finds

```
phv = [1, 1, 2, 2, 2, 4, 4, 4, 4]
nuv = [1, 2, 3, 4, 6, 5, 8, 10, 12]
```

5.4 Cyclotomic polynomials

Once we have these we essentially run `lineq` with input phv except that we also keep track of the corresponding cyclotomic polynomial, the product of the Φ_v's instead of just $\sum_{v \geq 1} \gamma_v \phi(v)$ (its degree).

The stack objects are triples s, pol, j, kept in three vectors sdeg, spol, sj, corresponding respectively to deg(P), P, j, where P is a cyclotomic polynomial, deg(P) is its degree and j is the index in the vector of v's nuv used to obtain P from the previous polynomial in the stack.

Here is a list of all cyclotomic polynomials of degree 4.

```
? p4v=polcyclolist(4);for(k=1,length(p4v),print(p4v[k]))

x^4 - 4*x^3 + 6*x^2 - 4*x + 1
x^4 - 2*x^3 + 2*x - 1
x^4 - 2*x^2 + 1
x^4 - x^3 - x + 1
x^4 - 2*x^3 + 2*x^2 - 2*x + 1
x^4 - 3*x^3 + 4*x^2 - 3*x + 1
x^4 + 2*x^3 - 2*x - 1
x^4 + x^3 - x - 1
x^4 - 1
x^4 - x^3 + x - 1
x^4 + 4*x^3 + 6*x^2 + 4*x + 1
x^4 + 3*x^3 + 4*x^2 + 3*x + 1
x^4 + 2*x^3 + 2*x^2 + 2*x + 1
x^4 + x^3 + x + 1
x^4 + 2*x^3 + 3*x^2 + 2*x + 1
x^4 + x^3 + 2*x^2 + x + 1
x^4 + x^2 + 1
x^4 + 2*x^2 + 1
x^4 - x^3 + 2*x^2 - x + 1
x^4 - 2*x^3 + 3*x^2 - 2*x + 1
x^4 + x^3 + x^2 + x + 1
x^4 + 1
x^4 - x^3 + x^2 - x + 1
x^4 - x^2 + 1
```

Again we should test our function; for example, like this

```
? ph10v=eulerphilist(10); 1/prod(n=1,5,
(1-x^(2*n))^length(ph10v[n])) + O(x^11),(1-x)^2)

1 + 2*x + 6*x^2 + 10*x^3 + 24*x^4 + 38*x^5 + 78*x^6 +
118*x^7 + 224*x^8 + 330*x^9 + 584*x^10 + O(x^11)
```

```
? sum(n=1,10,length(polcyclolist(n))*x^n,1+O(x^11))

1 + 2*x + 6*x^2 + 10*x^3 + 24*x^4 + 38*x^5 + 78*x^6 +
118*x^7 + 224*x^8 + 330*x^9 + 584*x^10 + O(x^11)
```

(and, again, why is this a test?)

5.5 Exercises

1. Change one line of part so that it computes the partitions $(\lambda_1,\ldots,\lambda_k)$ of n with at most r parts (i.e., with $k \leq r$).
2. Prove Euler's formula (5.5) for the generating function of $p(n)$.
3. Compare the running times for the different versions of partnum given in the text for small n (say $n = 1,\ldots,50$).
4. The following GP functions all compute the number of partitions $p(m)$ for $m \leq n$ using Euler's formula (5.5). Compare the running times of these functions for, say, $n = 50, 100, 200,\ldots$ and discuss their relative merits (if any).

   ```
   partnum1a(n) = Vec(1/prod(m=1,n,1-x^m,1+O(x^(n+1)))-1)

   partnum1b(n) = Vec(1/(prod(m=1,n,1-x^m)+O(x^(n+1)))-1)

   partnum1c(n) = Vec(1/prod(m=1,n,1-x^m)+O(x^(n+1))-1)

   partnum1d(n) = Vec(prod(m=1,n,1/(1-x^m+O(x^(n+1))))-1)

   partnum1e(n) = Vec(prod(m=1,n,1/(1-x^m))+O(x^(n+1))-1)
   ```

5. Analyze the algorithm in etaps and verify that it does compute the right hand side of (5.6) to precision $O(x^{n+1})$.
6. A *derangement* is a permutation with no fixed points. Let $d(n)$ be the number of derangments in S_n. Write a GP routine that computes $d(n)$.
7. Let d_n be the degree of the largest irreducible representation of S_n. Write a GP function that computes d_n for a given n (either using directly the routines of this chapter or, better yet, as a stand-alone routine in the style of part computing the values of f_λ along the way). Study numerically the behavior of $-\log(d_n/\sqrt{n!})/\sqrt{n}$.
8. Write a GP routine that computes the maximal order $g(n)$ of an element in S_n (it is known as Landau's function). The sequence $g(n)$ begins $1, 2, 3, 4, 6, 6, 12, 15, 20, 30, 30, 60,\ldots$. The sequence $1, 1, 2, 3, 4, 6, 6, 12,\ldots$ (i.e., with a tagged on 1 at the start) of values of $g(n)$ is Sloane A000793.
9. Verify numerically for small n the following formula of Frobenius and Hurwitz

$$\binom{n}{2}\frac{\chi^\lambda(C)}{\chi^\lambda(1)} = \sum_i \left[\binom{\lambda_i}{2} + (1-i)\lambda_i\right], \qquad (5.23)$$

where $\lambda = (\lambda_1, \lambda_2, \ldots)$ is a partition of n and C is the conjugacy class of a transposition in S_n.

10. Prove that $\prod_{|\lambda|=n} z_\lambda$, with z_λ is as in (5.12), is a square.
11. Using (5.23) write a GP routine that computes the number of solutions in S_n to the equation

$$\tau_1 \cdots \tau_m = 1 \tag{5.24}$$

where τ_i is a transposition.
12. A formula of Frobenius and Schur (valid for any finite group for which all irreducible representations are real, like S_n for example) gives

$$\sum_\lambda \chi^\lambda(w) = \#\{\sigma \in S_n \mid \sigma^2 = w\}, \tag{5.25}$$

for any $w \in S_n$. Write a GP routine implementing this formula. Check it for small n and $w = 1$ by verifying the identity

$$1 + \sum_{n \geq 1} \#\{\sigma \in S_n \mid \sigma^2 = 1\} \frac{x^n}{n!} = \exp\left(x + \frac{x^2}{2}\right) = 1 + x + \tfrac{1}{2}x^2 + \tfrac{2}{3}x^3 + \tfrac{5}{12}x^4 + \cdots \tag{5.26}$$

13. Combining (5.23) and (5.25) deduce that

$$\sum_\lambda \sum_i \left[\binom{\lambda_i}{2} + (1-i)\lambda_i\right] = 0, \tag{5.27}$$

where the sum is over all partitions $\lambda = (\lambda_1, \lambda_2, \ldots)$ of n. Write a GP routine to verify this formula for small n.
14. Let Γ_g be the fundamental group of a Riemann surface of genus $g > 0$. Given a finite group G a formula of Frobenius [85] gives

$$\#\operatorname{Hom}(\Gamma_g, G) = |G| \sum_\chi \left(\frac{|G|}{\chi(1)}\right)^{2g-2} \tag{5.28}$$

For $G = S_n$ this yields

$$\frac{1}{n!} \#\operatorname{Hom}(\Gamma_g, S_n) = \sum_\lambda h_\lambda^{2g-2}, \tag{5.29}$$

where the sum is over all partitions of n and h_λ is the product of all hook lengths of λ, see (5.7).

We may further combine this expression with (5.19). Write a GP routine that computes $u_n(\Gamma_g)$. Here is a short table for comparison.

g\n	1	2	3	4	5
1	1	3	4	7	6
2	1	15	220	5275	151086
3	1	63	7924	2757307	2081946006
4	1	255	281740	1542456475	29867372813886
5	1	1023	10095844	882442672507	429988374084026406

15. What would go wrong in `eulerphilist` if we changed the line `while (p <= n,` to `while (p < n,`? (Try it out!)
16. Modify the function `eulerphilist` so that it computes all values of $v \in \mathbf{N}$ with $f(v) \le n$ for other multiplicative functions, for example, the sum of divisors function $f(v) = \sum_{d|v} d$.
17. In the spirit of the algorithms of this chapter write a function that for given a, b, n outputs all strings of integers $x_1, \ldots, x_m$ with $x_i \le x_{i+1}$, $a \le x_i \le b$ and $m \le n$.

6 *p*-adic numbers

We will assume the reader is familiar with the basic concepts and notation of *p*-adic numbers; see, for example, [20], [82] or H. Cohen's new book [23].

For $x \in \mathbf{Q}_p$ we let $v_p(x)$ be (as before) its *valuation at p* and let

$$|x|_p := p^{-v_p(x)} \tag{6.1}$$

be its *p-adic absolute value* (normalized in the standard way).

6.1 Basic functions

6.1.1 Mahler's expansion

A continuous function $f : \mathbf{Z}_p \to \mathbf{Z}_p$ is uniquely determined by its values $f(n)$ at integers $n \in \mathbf{Z}_p$ as any *p*-adic integer is a limit of rational integers. A precise form of this fact is given by the following theorem due to Mahler ([20], p. 307).

Theorem 6.1 *A function* $f : \mathbf{Z}_p \to \mathbf{Z}_p$ *is continuous if and only if f is of the form*

$$f(x) = \sum_{n \geq 0} a_n \binom{x}{n}, \quad a_n \in \mathbf{Z}_p, \quad \lim_{n \to \infty} a_n = 0. \tag{6.2}$$

The expansion (6.2) is called the *Mahler expansion* of f. The coefficients a_n are uniquely determined by (the values of) f and satisfy

$$a_n = \sum_{k=0}^{n} (-1)^{n-k} \binom{n}{k} f(k), \quad f(n) = \sum_{k=0}^{n} \binom{n}{k} a_k. \tag{6.3}$$

These relations can be conveniently condensed in the following identity of formal power series in a variable T.

$$e^{-T} \sum_{n \geq 0} f(n) \frac{T^n}{n!} = \sum_{n \geq 0} a_n \frac{T^n}{n!}. \tag{6.4}$$

An important example of a Mahler expansion is when $a_n = z^n$ for a fixed $z \in p\mathbf{Z}_p$ for which we get

$$(1+z)^x = \sum_{n \geq 0} z^n \binom{x}{n}, \quad x \in \mathbf{Z}_p, \; z \in p\mathbf{Z}_p. \tag{6.5}$$

In particular, for an integer $n \equiv 1 \bmod p$ and $x \in \mathbf{Z}_p$ we may use (6.5) to define the value of n^x.

6.1.2 Hensel's lemma and Newton's method (again)

The same methods used in §4.1.4 apply almost verbatim to the p-adic situation replacing the valuation v_x by the p-adic valuation v_p. For example, suppose $f \in \mathbf{Z}_p[x]$ and we have a solution $f(x_0) \equiv 0 \bmod p$ for some $x_0 \in \mathbf{Z}_p$ with $f'(x_0)$ not $0 \bmod p$. Then we can construct recursively a sequence $x_n \in \mathbf{Z}_p$ such that $x_n \equiv x_{n-1} \bmod p^n$ and $f(x_n) \equiv 0 \bmod p^n$. In other words, the solution $x_0 \bmod p$ lifts to a p-adic solution $\lim_n x_n$.

As a simple example, suppose we want to solve the equation $f(x) := x^{11} + x - a = 0$ with $a \in \mathbf{Z}_{11}$. Considered modulo 11 we have $2x \equiv a \bmod 11$ which has only one solution $x_0 \equiv 6a \bmod 11$. Since $f'(x) \bmod 11$ is identically $1 \neq 0$ this solution extends to a unique $\mathbf{Z}_{11}$ solution: the limit of the sequence defined by the following recursion

$$x_0 := 2a, \qquad x_n = a - x_{n-1}^{11}, \quad n \geq 1.$$

For $a = 1$, say, we obtain

$$x = 6 + 2 \cdot 11 + 10 \cdot 11^3 + 6 \cdot 11^4 + 4 \cdot 11^5 + 2 \cdot 11^6$$
$$+ 4 \cdot 11^7 + 9 \cdot 11^8 + 9 \cdot 11^9 + O(11^{10}).$$

As a check we can do

```
? polrootspadic(x^11+x-1,11,10)

[6 + 2*11 + 10*11^3 + 6*11^4 + 4*11^5 + 2*11^6 + 4*11^7 +
9*11^8 + 9*11^9 + O(11^10)]~
```

Loosely speaking then, if we stay away from singularities, solving equations p-adically is the same as solving them modulo p and the p-adic solution can be constructed from the modulo p solution recursively. (In general, we need a solution to a high enough power of p, depending on the nature of the singularity, before the recursion kicks in.)

6.1 Basic functions

To illustrate this further let us return to the square root algorithm newtsqrt which applied Newton's method to the equation $x^2 - D$. That is, consider the recursion(4.22). Assume $D \in 1 + p\mathbf{Z}_p$ and $p > 2$. Then D is a square in $\mathbf{Z}_p$ and the *exact* same recursion will compute the square root of D in $1 + p\mathbf{Z}_p$. In the examples below z is a random choice.

```
? z=1+5/3+O(5^10);newtsqrt(z,3,1,0)-sqrt(z)

2*5^8 + 3*5^9 + O(5^10)
? z=1+5/3+O(5^20);newtsqrt(z,4,1,0)-sqrt(z)

2*5^16 + 4*5^17 + 2*5^19 + O(5^20)
? z=1+5/3+O(5^36);newtsqrt(z,5,1,0)-sqrt(z)

2*5^32 + 5^33 + 5^34 + 4*5^35 + O(5^36)
```

Here we have tacitly used the fact that the GP built-in function sqrt maps $1 + p\mathbf{Z}_p$ to itself.

For $p = 2$ the equation $x^2 - D$ has derivative identically zero mod 2. Hence Hensel's lemma as we stated it does not apply directly. As we mentioned, we need to start with a sufficiently good approximation to get the recursion going and lift the solution to a 2-adic one. For the square root case it is enough to have a solution $x_0^2 \equiv D$ mod 8.

```
? z=1+8/9+O(2^20);newtsqrt(z,4,1,0)+sqrt(z)

2^17 + O(2^19)
? z=1+8/9+O(2^36);newtsqrt(z,5,1,0)+sqrt(z)

2^33 + O(2^35)
? z=1+8/9+O(2^70);newtsqrt(z,6,1,0)+sqrt(z)

2^65 + 2^67 + 2^68 + O(2^69)
```

Note that for $p = 2$ the function sqrt does not map $1 + 8\mathbf{Z}_2$ to itself! It is always dangerous to assume that functions will compute what we think is the natural answer when there is a choice.

Another interesting case where we have a recursion converging to a fixed point is in the definition of the Teichmüller character (see below). The equation in question is now $x^p = x$ which has derivative identically 1 modulo p, and hence non-zero, in all cases.

For any $x \in \mathbf{Z}_p$ the sequence $x, x^p, x^{p^2}, \ldots$ in $\mathbf{Z}_p$ converges. The limit ζ naturally satisfies $\zeta^p = \zeta$ and $\zeta \equiv x \bmod p$. In an example we get

```
? z=23/17+O(7^10);w=teichmuller(z);
? z^7-w

4*7^2 + 7^3 + 7^4 + 6*7^5 + 6*7^6 + 2*7^7 + 7^8 + 7^9 + O(7^10)

? z^(7^2)-w

4*7^3 + 7^4 + 3*7^5 + 4*7^6 + 5*7^9 + O(7^10)

? z^(7^3)-w

4*7^4 + 7^5 + 3*7^6 + 6*7^7 + 5*7^8 + 3*7^9 + O(7^10)
```

So we see that the convergence is linear at best. If we use Newton's method we get the recursion

$$x_{n+1} := x_n - \frac{x_n^p - x_n}{p x_n^{p-1} - 1}, \quad x \in \mathbf{Z}_p \tag{6.6}$$

(the function on the right is $x - f(x)/f'(x)$ with $f(x) = x^p - x$). This, as expected, works much better.

```
? F(x,p)=x-(x^p-x)/(p*x^(p-1)-1)
? z=23/17+O(7^20);w=teichmuller(z);
? z=F(z,7);z-w

5*7^3 + 3*7^4 + 4*7^5 + 3*7^6 + 3*7^7 + 5*7^9 + 2*7^12 +
7^13 + 4*7^14 + 2*7^15 + 2*7^16 + 7^17 + 5*7^18 + 3*7^19 +
O(7^20)

? z=F(z,7);z-w

3*7^7 + 2*7^10 + 2*7^11 + 6*7^12 + 7^13 + 6*7^14 + 2*7^15 +
2*7^16 + 7^17 + 5*7^18 + 3*7^19 + O(7^20)

? z=F(z,7);z-w

5*7^15 + 7^16 + 5*7^17 + 7^18 + 5*7^19 + O(7^20)
```

To $a \in (\mathbf{Z}/p\mathbf{Z})^\times$ we may associate its *Teichmüller representative*, $\zeta = \lim_{n \to \infty} x^{p^n}$, where $x \in \mathbf{Z} \subseteq \mathbf{Z}_p$ is such that $a = x \bmod p$.

This gives rise to the *Teichmüller character*

$$\text{Teich}: \mathbf{F}_p^\times \longrightarrow \mathbf{Z}_p^\times \qquad (6.7)$$
$$a \mapsto \zeta,$$

which gives an isomorphism between $\mathbf{F}_p^\times$ and the $(p-1)$-th roots of unity in $\mathbf{Z}_p$ (the only roots of unity it has if $p > 2$).

6.2 The p-adic gamma function

The (Morita) p-adic gamma function Γ_p (see [82], p. 176) is the unique continuous function $\mathbf{Z}_p \longrightarrow \mathbf{Z}_p$ which interpolates the following values at natural numbers $n \in \mathbf{N}$

$$\Gamma_p(n) = (-1)^n \prod_{\substack{k=1 \\ p \nmid k}}^{n-1} k. \qquad (6.8)$$

In particular, $\Gamma_p(0) = 1$ (by definition an empty product equals 1). Since Γ_p is continuous we deduce that $\Gamma_p(x) \in \mathbf{Z}_p^\times$ for all $x \in \mathbf{Z}_p$. It satisfies several identities analogous to the complex Γ function (see §6.88). For example,

$$\Gamma_p(x+1) = -\Gamma_p(x) \begin{cases} 1 & x \in p\mathbf{Z}_p \\ x & \text{otherwise} \end{cases} \qquad (6.9)$$

and [78]

$$\Gamma_p(x)\Gamma_p(1-x) = (-1)^{1+y_0+y_1(p-1)}, \qquad (6.10)$$

where

$$-x = y_0 + y_1 p + O(p^2), \quad 0 \le y_0, y_1 < p. \qquad (6.11)$$

A further property that we will need is the following [82], Prop. 35.3

$$|\Gamma_p(x) - \Gamma_p(y)|_p \le |x-y|_p, \quad p > 3. \qquad (6.12)$$

In order to compute $\Gamma_p(x)$ efficiently for an arbitrary $x \in \mathbf{Z}_p$ we will use the following Mahler expansion.

Theorem 6.2 *Let p be a prime number. Define rational numbers c_n by the power series expansion*

$$\exp\left(x + \frac{x^p}{p}\right) =: \sum_{n \ge 0} c_n x^n. \qquad (6.13)$$

6 : p-adic numbers

Then for $0 \le a < p$ and $x \in \mathbf{Z}_p$

$$\Gamma_p(-a+px) = \sum_{k\ge 0} p^k c_{a+kp}(x)_k, \tag{6.14}$$

where

$$(x)_k := x(x+1)\cdots(x+k-1). \tag{6.15}$$

Note that $\binom{-x}{k} = (-1)^k (x)_k/k!$ so (6.14) is essentially a Mahler expansion.

It is actually more convenient to modify the coefficients c_n and consider instead the numbers defined by

$$d_{a+kp} := p^k k!\, c_{a+kp}, \quad 0 \le a < p,\ k \in \mathbf{N} \tag{6.16}$$

so that we have

$$\Gamma_p(-a+px) = \sum_{k\ge 0} d_{a+kp} \frac{(x)_k}{k!}. \tag{6.17}$$

Differentiating both sides of (6.13) it is easy to see that the c_n satisfy the recursion

$$nc_n = c_{n-1} + c_{n-p}, \quad c_0 = 1, \quad c_n = 0,\ \text{for } n < 0. \tag{6.18}$$

The corresponding recursion for d_n is

$$nd_n = d_{n-1} + p\left\lfloor \frac{n}{p} \right\rfloor d_{n-p}, \quad \text{if } p \nmid n \tag{6.19}$$

and otherwise

$$d_{kp} = d_{kp-1} + d_{(k-1)p}, \quad k \in \mathbf{N}, \tag{6.20}$$

with the same initial conditions $d_0 = 1$ and $d_n = 0$ for $n < 0$. Note that it is now clear that $d_n \in \mathbf{Z}_p$.

These recursions give a convenient way to compute c_n or d_n keeping in memory only p of them at any given time. Here is our implementation of (6.17) in GP.

```
gammap(x) =
{
  local(n,a,p,m,d,t,s);
  p=component(x,1);
  m=padicprec(x,p);
  n=ceil(p*m/(p-1));
  a=-x%p;
  x=(x+a)/p;a++;
  x=truncate(x)+O(p^(m+floor(n/(p-1))));
  d=vector(p);
```

```
  d[1]=1+O(p^m);t=1;
    for(j=1,p-1,d[j+1]=d[j]/j);
    s=d[a];
      for(k=1,n-1,
        d[1]=d[1]+d[p];
        for(j=1,p-1,d[j+1]=(d[j]+p*k*d[j+1])/(k*p+j));
        t*=(x+k-1)/k;
        s+=d[a]*t);
  s
}
```

The argument x should be a p-adic number. We recover p from x by `p=component(x,1)` and `m=padicprec(x,p)` gives us the precision of x.

To compute the number of terms to take in (6.17) we use the estimate

$$v_p(d_{a+kp}) \geq k - \left\lfloor \frac{k}{p} \right\rfloor, \quad 0 \leq a < p. \tag{6.21}$$

which follows from work of Dwork (see for example [78], p. 167). In particular, $v_p(d_{a+kp})$ grows at least linearly with k (in fact, the estimate (6.21) is quite sharp).

It is not hard to verify then that if we take

$$n := \left\lceil \frac{p}{p-1} m \right\rceil \tag{6.22}$$

then $n - \lfloor n/p \rfloor \geq m$ (see Ex. 5). Since $(x)_k/k! \in \mathbb{Z}_p$ for $x \in \mathbb{Z}_p$, truncating the series (6.17) at n gives, by the non-Archimedean principle, $\Gamma_p(x)$ with precision $O(p^m)$. An important point to make, however, is that division by k in `t*=(x+k-1)/k` (to compute $(x)_k/k!$ recursively) could potentially yield an overall loss of precision in the terms of the sum. This loss is at most $v_p(n!)$, which is smaller than $n/(p-1)$ (see Ex. 3). To avoid this problem we increase the precision with which we compute $(x)_k/k!$ by setting

```
x=truncate(x)+O(p^(m+floor(n/(p-1))))}.
```

By a simple modification of this routine we can compute all values of $\Gamma_p(-a+px)$ with $0 \leq a < p$ and fixed $x \in \mathbb{Z}_p$ simultaneously.

```
gammapv(x)=
{
  local(n,a,p,m,d,t,v);
  p=component(x,1);
  m=padicprec(x,p);
  n=ceil(p*m/(p-1));
  x=truncate(x)+O(p^(m+floor(n/(p-1))));
```

```
d=vector(p);v=d;
d[1]=1+O(p^m);v[1]=d[1];t=1;
  for(j=1,p-1,d[j+1]=d[j]/j);
  for(j=2,p,v[j]=d[j]);
    for(k=1,n-1,
      t*=(x+k-1)/k;
      d[1]=d[1]+d[p]; v[1]+=d[1]*t;
      for(j=1,p-1,d[j+1]=(d[j]+p*k*d[j+1])/(k*p+j));
      for(j=2,p,v[j]+=d[j]*t));
v
}
```

The output is a vector v of length p with $v[k] = \Gamma_p(1 - k + px)$ for $k = 1, 2, \ldots, p$.

6.2.1 The multiplication formula

The classical gamma function (6.88) satisfies for any $n \in \mathbf{N}$ the following multiplication formula

$$\Gamma\left(\frac{x}{n}\right)\Gamma\left(\frac{x+1}{n}\right)\cdots\Gamma\left(\frac{x+n-1}{n}\right) = \frac{(2\pi)^{\frac{1}{2}(n-1)}}{n^{x-\frac{1}{2}}}\Gamma(x). \tag{6.23}$$

The corresponding formula [78], p. 371 for Γ_p is as follows: for $p \nmid n$

$$\Gamma_p\left(\frac{x}{n}\right)\Gamma_p\left(\frac{x+1}{n}\right)\cdots\Gamma_p\left(\frac{x+n-1}{n}\right) = \frac{\omega}{n^c}\Gamma_p(x), \tag{6.24}$$

where $\omega = \prod_{k=1}^{n-1} \Gamma_p\left(\frac{k}{n}\right)$ depends only on p and n,

$$\omega^2 = \begin{cases} -1 & \text{if } p \equiv 1 \bmod 4 \text{ and } n \text{ is even} \\ 1 & \text{otherwise} \end{cases} \tag{6.25}$$

and

$$c = z_0 + (p-1)(z_1 + pz_2 + \cdots)$$

with

$$x - 1 = z_0 + z_1 p + z_2 p^2 + \cdots, \quad 0 \le z_i < p.$$

(Here the term n^c on the right-hand side of (6.24) is understood as $n^{z_0} \cdot (n^{p-1})^{z_1+pz_2+\cdots}$.)

We can be more precise about the value of ω for n odd.

Proposition 6.3 For any prime p not diving n we have

$$(-1)^{\frac{1}{2}(n-1)} \prod_{k=1}^{n-1} \Gamma_p\left(\frac{k}{n}\right) = \left(\frac{p}{n}\right), \quad n \text{ odd}. \tag{6.26}$$

Proof Assume first that $p > 2$. Let $Y = \{1, 2, \ldots, (n-1)/2\}$. For each $y \in Y$ write

$$2py = a + nx, \quad 0 < a < n, \quad 0 \le x < p. \tag{6.27}$$

Let A be the set of resulting values of a as y runs over Y. It is clear that A is set of representatives for the involution $a \mapsto n-a$ on the set $\{1, 2, \ldots, n-1\}$. Clearly

$$x = \left\lfloor \frac{2py}{n} \right\rfloor \tag{6.28}$$

and

$$-\frac{a}{n} \equiv x \bmod p. \tag{6.29}$$

Pairing the factors a/n and $1 - a/n$ in the product (6.26) we have, using (6.10), that

$$(-1)^{(n-1)/2} \prod_{k=1}^{n-1} \Gamma_p\left(\frac{k}{n}\right) = \prod_{a \in A} (-1)^x. \tag{6.30}$$

To finish the proof we appeal to the following general version of Gauss's lemma

$$\prod_{y=1}^{\frac{1}{2}(n-1)} (-1)^{\left\lfloor \frac{2zy}{n} \right\rfloor} = \left(\frac{z}{n}\right) \tag{6.31}$$

valid for any $z \in \mathbf{N}$ coprime to n, n odd, where the right-hand side denotes the Jacobi symbol.

The proof for $p = 2$ is along the same lines; instead of the factor $(-1)^x$ we now have $(-1)^{x+[x/2]}$ and we use (6.31) for $z = 1$ and $z = 2$. We leave the details to the reader. $\square$

Remark The above proposition appears to be new (see also [23]). It will not come as a surprise to the reader that its general form was discovered by computing the left-hand side of (6.26) for several primes p for a fixed n and then using the function testmod of chapter 1 to test the modularity of the resulting sequence of ± 1.

There is an analogous formula for n even (see [23]) where the right-hand side also has a factor of $\Gamma_p(\frac{1}{2})^{n-1}$; however, $\Gamma_p(\frac{1}{2})$ varies with p in a seemingly random way and a closed formula seems unlikely (see Ex. 7).

6.3 The logarithmic derivative of Γ_p

A function is called *locally analytic* if it can be expressed as a convergent power series on a disk about every point of its domain.

The function Γ_p is locally analytic (see [82] thm. 58.4). We define

$$\psi_p(x) := \frac{\Gamma'_p(x)}{\Gamma_p(x)}. \tag{6.32}$$

As we observed above $\Gamma_p(x) \in \mathbf{Z}_p^\times$ for all $x \in \mathbf{Z}_p$ hence ψ_p defines a continuous function from $\mathbf{Z}_p$ to $\mathbf{Z}_p$. It follows from (6.9) that

$$\psi_p(x+1) - \psi_p(x) = \begin{cases} 0 & x \in p\mathbf{Z}_p \\ x^{-1} & \text{otherwise.} \end{cases} \tag{6.33}$$

We can easily modify our routine for Γ_p and write one that computes ψ_p by differentiating term by term the series (6.17). We then use the recursion

$$(x)'_k = (x)_{k-1} + (x+k-1)(x)'_{k-1} \tag{6.34}$$

obtained by differentiating the corresponding recursion for $(x)_k$. There will be a bit more of precision loss now so we need to account for this. Note that

$$(x)'_k = (x)_k \sum_{j=0}^{k-1} \frac{1}{x+j}. \tag{6.35}$$

It follows that we lose at worst $\lceil \log_p(n) \rceil$ powers of p if we use n terms in the sum ($\log_p(x)$ is the logarithm of the real number x in base p not the p-adic logarithm). As we did with Γ_p what we need is

$$n - \left\lceil \frac{n}{p} \right\rceil - \lceil \log_p(n) \rceil \geq m \tag{6.36}$$

to obtain precision $O(p^m)$. In the following implementation we first find n as we did for gammap and then increase n successively until (6.36) is satisfied. There will also be some precision loss in computing the terms $(x)'_k$ themselves and hence we also increase the precision of x accordingly.

To avoid using real numbers altogether we use a simple function clogp (for *ceiling of* $\log_p$) to compute $\lceil \log_p(\cdot) \rceil$.

```
clogp(x,p)=
{
  local(k);
  x=abs(x);
  k=0;
    until(x==0,x=x\p;k++);
  k
}
```

(See Ex. 9 for an alternative.)

6.3 The logarithmic derivative of Γ_p 161

Here is finally our GP implementation of `psip`; `t`, `t1` keep track of $(x)_k$ and $(x)'_k$ respectively and `s`, `s1` are the corresponding sums approximating $\Gamma_p(x)$ and $\Gamma'_p(x)$.

```
psip(x)=
{
  local(n,a,p,pp,m,np,d,t,t1,s,s1);
  p=component(x,1);
  m=padicprec(x,p);
  n=ceil(p*m/(p-1));
  np=clogp(n,p);
  while(n-floor(n/p)-np < m,
    n++;
    np=clogp(n,p));
  m+=np;
  a=-x%p;
  x=(x+a)/p;a++;
  x=truncate(x)+O(p^(m+floor(n/(p-1))));
  d=vector(p);
  d[1]=1+O(p^m);t=1;t1=0;
  for(j=1,p-1,d[j+1]=d[j]/j);
  s=d[a];s1=0;
  for(k=1,n-1,
    d[1]=d[1]+d[p];
    for(j=1,p-1,d[j+1]=(d[j]+p*k*d[j+1])/(k*p+j));
    t1*=(x+k-1);t1+=t;t1/=k;
    t*=(x+k-1)/k;
    s+=d[a]*t;
    s1+=d[a]*t1);
  s1/s/p
}
```

6.3.1 Application to harmonic sums

It follows from (6.33) that for $n \in \mathbf{N}$

$$\psi_p(n+1) - \psi_p(1) = \sum_{\substack{k=1 \\ p \nmid k}}^{n} \frac{1}{k}, \qquad (6.37)$$

and if we let

$$H_n := \sum_{k=1}^{n} \frac{1}{k}, \quad n \in \mathbf{N} \qquad (6.38)$$

be the harmonic sum then

$$H_n = \sum_{k \geq 0} p^{-k} \left(\psi_p \left(\left\lfloor \frac{n}{p^k} \right\rfloor + 1 \right) - \psi_p(1) \right). \tag{6.39}$$

This formula is now easy to implement; in the following version the input is n,p,m and the output is $H_n + O(p^m)$ (m is actually optional and set arbitrarily to 20 if not given).

```
harmp(n, p, m = 20) =
{
  local(psip1, s = 0,pp = 1);
  m+=clogp(n,p);
  psip1=psip(1+O(p^m));

    while(n,
           s += (psip(n+1+O(p^m)) - psip1) / pp;
           n\=p; pp*=p);
  s
}
```

Note that we increase the working precision by $\lceil \log_p(n) \rceil$ to account for the factors p^{-k} in (6.39).

The function harmp only becomes useful when n is much larger than p since otherwise using the definition of H_n, say,

```
harmp1(n,p,m)=sum(k=1,n,1/k,O(p^m))
```

would require fewer operations.

As an example, let us verify the following fact discovered by Boyd [15]: $v_{11}(H_n) = 3$ for $n = 848, 9338, 10583$ and 3546471722268916272.

```
? v=[848,9338,10583,3546471722268916272];
? for(k=1,4,print(harmp(v[k],11,8)))

9*11^3 + 8*11^4 + 9*11^5 + 9*11^6 + 10*11^7 + 9*11^8 + O(11^9)
2*11^3 + 2*11^4 + 9*11^5 + 3*11^6 + 9*11^7 + 11^8 + O(11^9)
11^3 + 2*11^5 + 7*11^6 + 3*11^7 + 7*11^8 + O(11^9)
5*11^3 + 4*11^6 + 10*11^8 + O(11^9)
```

6.3.2 A formula of J. Diamond

Let d be a divisor of $p - 1$. Then in [28], p. 336, Diamond proved the following analogue of a theorem of Gauss

$$\psi_p\left(\frac{r}{d}\right) = \psi_p(0)$$

$$+ \left(1 - \frac{1}{p}\right)\left(-\log d + \sum_{k=1}^{d-1} \zeta_d^{-kr}\log(1 - \zeta_d^k)\right), \quad 0 \leq r < d,$$

(6.40)

where $\zeta_d \in \mathbb{Z}_p$ is any primitive d-th root of unity. Here $\psi_p(0) = \psi_p(1)$ appears as a p-adic analogue of Euler's constant ($\gamma = -\Gamma'(1) = 0.5772156649\cdots$) and can be obtained also as

$$\psi_p(0) = \lim_{k \to \infty} p^{-k} \sum_{\substack{n=1 \\ p \nmid n}}^{p^k - 1} \log(n).$$

(6.41)

Combined with Dirichlet's class number formula for a real quadratic field F (6.40) implies that

$$\sum_r \left(\frac{d}{r}\right) \psi_p\left(\frac{r}{d}\right) = -\sqrt{d}\left(1 - \frac{1}{p}\right) 2h \log \epsilon,$$

(6.42)

where $d = \text{disc}(F)$, h is the class number of F and $\epsilon > 1$ is a generator of the unit group of F up to torsion.

To test this formula numerically we can try this:

```
diam(d,p, m = 10) =
{
  -sum(r=1,d-1,
    kronecker(d,r)*psip(r/d+O(p^m)))
    /sqrt(d+O(p^m))/(1-1/p)/2
}
```

(In the next function we assume the input d is a fundamental positive discriminant.)

```
logeps(d,p,m = 10) = log(quadunit(d) + O(p^m))
```

We can test the formula now with say

```
diamtest(p,Dmax,m) =
{
  for(d=2,Dmax,
    if(issquarefree(d),
      D=quaddisc(d);
      if(D<Dmax && kronecker(D,p)==1,
      print(D," ",diam(D,p,m)/
      logeps(D,p,m)/qfbclassno(D)))))
}
```

For example,

```
? diamtest(13,100,20)

12   1 + O(13^19)
40   1 + O(13^19)
56   1 + O(13^19)
17   1 + O(13^19)
88   1 + O(13^19)
92   1 + O(13^19)
29   1 + O(13^19)
53   1 + O(13^19)
61   1 + O(13^19)
69   1 + O(13^19)
77   1 + O(13^19)
```

The attentive reader will have noticed that we are actually testing formula (6.42) in far more cases than those claimed (we are only requiring that p splits in F and not that $d \mid p - 1$; see also Ex. 10). For details on the more general formula see [23].

6.3.3 Power series expansion of $\psi_p(x)$

In the disk $|x|_p < 1$ the function $\psi_p(x)$ has the power series expansion

$$\psi_p(x) = b_0 - \sum_{n \geq 1} \frac{b_n}{n} x^n, \quad |x|_p < 1, \tag{6.43}$$

with

$$b_0 = \int_{\mathbb{Z}_p^\times} \log x \, dx, \quad b_n = \int_{\mathbb{Z}_p^\times} x^{-n} \, dx. \tag{6.44}$$

Here $\int_{\mathbb{Z}_p^\times}$ stands for the so-called *Volkenborn integral* defined by

$$\int_{\mathbb{Z}_p} f(x) \, dx := \lim_{n \to \infty} p^{-n} \sum_{k=0}^{p^n - 1} f(k) \tag{6.45}$$

6.3 The logarithmic derivative of Γ_p

and for a compact set $U \subseteq \mathbf{Z}_p$ by

$$\int_U f(x)\, dx := \int_{\mathbf{Z}_p} \chi_U(x) f(x)\, dx,$$

where χ_U is the characteristic function of U. (See [82], §55, 57; compare with §1.3.3. For a proof of (6.43), differentiate the power series for $\log \Gamma_p$ given in [82], Lemma 58.1.) Note that $b_0 = \psi_p(0)$ and the integral expression for b_0 is then equivalent to (6.41).

The coefficients b_n can be expressed in terms of Bernoulli numbers. Indeed, we have

$$\int_{\mathbf{Z}_p} x^n\, dx = B_n, \qquad (6.46)$$

with B_n the n-th Bernoulli number (Ex. 12). Hence,

$$\int_{\mathbf{Z}_p^\times} x^n\, dx = (1 - p^{n-1}) B_n. \qquad (6.47)$$

We can write the integral for b_n, $n > 0$ as

$$\int_{\mathbf{Z}_p^\times} x^{-n}\, dx = \lim_{s \to \infty} \int_{\mathbf{Z}_p^\times} x^{(p-1)p^s - n}\, dx \qquad (6.48)$$

and conclude that

$$b_n = \lim_{s \to \infty} B_{[-n+(p-1)p^s]} \qquad n > 0. \qquad (6.49)$$

For example, using the function `bernoulli` from §1.3.2 (or simply the GP built-in version `bernfrac`) we may check (Ex. 13) that for $p = 3$

```
b_2 = 2*3^-1 + 2 + 3 + 2*3^2 + 2*3^3 + O(3^5)
b_4 = 2*3^-1 + 2 + 3 + 3^3 + 2*3^4 + O(3^5)
b_6 = 2*3^-1 + 3^3 + 2*3^4 + O(3^5)
b_8 = 2*3^-1 + 2 + 2*3 + 3^2 + 2*3^4 + O(3^5)
```

(Compare this discussion with [15] §5.2.)

In particular, $b_n = 0$ for n odd and, by the von Staudt–Clausen theorem (see [62], p. 291), b_n is p integral unless $p - 1 \mid n$, in which case it has denominator p. It follows easily that

$$v_p(\psi_p(p^k) - \psi_p(0)) \geq 2k + v_p\left(\frac{b_2}{2}\right) \qquad (6.50)$$

where

$$v_p\left(\frac{b_2}{2}\right) = \begin{cases} -2 & \text{if } p = 2 \\ -1 & \text{if } p = 3 \end{cases} \qquad (6.51)$$

and
$$v_p\left(\frac{b_2}{2}\right) \geq 0 \quad \text{if } p > 3. \tag{6.52}$$

For $k = 1$ and $p > 3$ this gives the following theorem of Wolstenholme [46], p. 89.
$$H_{p-1} \equiv 0 \bmod p^2, \quad p > 3. \tag{6.53}$$

Remark Equality holds in (6.52) almost all the time; in fact, there are only two known primes for which it fails, namely, $p = 16843$ and $p = 2124679$, where $p \mid B_{p-3}$. See the comments in [15] §4.

We can test this by modifying the routine bernoulli of Chapter 4 for p-adic inputs. As usual, we increase the precision appropriately to account for the division performed (a loss of not more than $v_p((n+1)!)$ on each of the variables c and s). But we now encounter a new problem. In computing x^n if x is divisible by p and n is large we will encounter an overflow error with GP. To avoid this we ignore this term altogether if $v_p(x)n$ is bigger than the working precision m1. (The problem would still surface if both n and the desired precision are large but we will assume this is not the case.) Here is the resulting routine.

```
bernoullip(n,x)=
{
  local(h,s,c,p,m,m1,r);

  p=component(x,1);
  m=padicprec(x,p);
  m1=m+floor((n+1)/(p-1));
  x=truncate(x)+O(p^(m1));

  h=O(p^m);s=O(p^m1);c=-1+O(p^(m1));

  for(k=1,n+1,
    c*=1-(n+2)/k;
    r=valuation(x,p);
    s+=if(r*n<m1,x^n,0);
    x++;
    h+=c*s/k);
  h
}
```

We can now perform our check.

```
? bernoullip(16843-3,O(16843^2))

 7786*16843 + O(16843^2)
```

```
? bernoullip(2124679-3,O(2124679^2))

1390423*2124679 + O(2124679^2)
```

The last calculation took a few minutes in a desktop machine. Computing the actual Bernoulli number would be pretty hard: by (1.51) $\log |B_{2124676}| \sim 5.27 \times 10^8$.

6.3.4 Application to congruences

We may also deduce other known congruences from (6.43). For example, in Ex. 6 in chapter I of [88] asks us to prove that

$$\binom{pa}{pb} \equiv \binom{a}{b} \mod p^3, \quad p > 3. \tag{6.54}$$

Consider more generally

$$A(n) := \prod_{\nu \geq 1} (\nu n)!^{\gamma_\nu}, \tag{6.55}$$

where $\gamma_\nu \in \mathbf{Z}$ is zero except finitely many and

$$\sum_{\nu \geq 1} \nu \gamma_\nu = 0. \tag{6.56}$$

From the definition (6.8) we obtain

$$\Gamma_p(n+1) = (-1)^{n+1} \frac{n!}{p^m m!}, \quad m = \left\lfloor \frac{n}{p} \right\rfloor \tag{6.57}$$

Therefore, under our assumption (6.56),

$$\frac{A(p)}{A(1)} = (-1)^{\sum_{\nu \geq 1} \gamma_\nu} \prod_{\nu \geq 1} \Gamma_p(\nu p + 1)^{\gamma_\nu} \tag{6.58}$$

and by the functional equation (6.9)

$$\frac{A(p)}{A(1)} = \prod_{\nu \geq 1} \Gamma_p(\nu p)^{\gamma_\nu}. \tag{6.59}$$

Taking p-adic logarithms on both sides we obtain

$$\log(A(p)/A(1)) = \sum_{\nu \geq 1} \gamma_\nu \log \Gamma_p(\nu p). \tag{6.60}$$

Now using the power series expansion

$$\log \Gamma_p(x) = b_0 x - \sum_{m=1}^{\infty} b_m \frac{x^{m+1}}{m(m+1)}, \quad |x|_p < 1, \tag{6.61}$$

with b_m as in (6.43) ([82] lemma 58.1, compare with (4.37)) we obtain

$$\log(A(p)/A(1)) = -\sum_{m=1}^{\infty} b_m \frac{p^{m+1}}{m(m+1)} \sum_{\nu \geq 1} \gamma_\nu \nu^{m+1} \qquad (6.62)$$

again thanks to our assumption (6.56). The terms in the sum have valuation at least 3 since $v_p(b_2/6) \geq 0$ for $p > 3$ and $b_n = 0$ for n odd. We conclude that

$$\log(A(p)/A(1)) \equiv 0 \bmod p^3, \quad p > 3. \qquad (6.63)$$

It follows from (6.12) that $\Gamma_p(px) \equiv 1 \bmod p$ for $x \in \mathbf{Z}_p$ and hence exponentiating (6.63) we find

$$\frac{A(p)}{A(1)} \equiv 1 \bmod p^3, \quad p > 3, \qquad (6.64)$$

which is a stronger congruence than (6.54). By an analogous argument we find

$$\frac{A(p^s)}{A(p^{s-1})} \equiv 1 \bmod p^{3s}, \quad p > 3, \quad s \in \mathbf{N}, \qquad (6.65)$$

and in particular $\lim_{s \to \infty} A(p^s)$ exists (see Ex. 16).

It is now easy to create examples with congruences like (6.54) valid to a higher power of p. For example,

$$A(pn) \equiv A(n) \bmod p^5, \quad A(n) := \frac{(2n)!^4}{n!^5(3n)!}, \quad p > 5. \qquad (6.66)$$

For more congruences of a similar kind see Ex. 17 and [94].

6.4 Analytic continuation

Let $f = \sum_{n \geq 0} a_n x^n$ be a power series with $a_n \in \mathbf{Z}_p$ convergent in the disk $D^- := \{x \mid |x|_p < 1\}$ about the origin. We would like to know if f can be "analytically continued" to a larger domain. In contrast to what happens over $\mathbf{C}$, however, we cannot simply expand f in power series about other points of D^-; the region of convergence of any such expansion will also be D^- (in particular, f is locally analytic in D^-).

Krasner discovered that one may use an alternative method due to Runge: consider uniform limits of rational functions with poles in the complement of the domain in question. The details are too technical and beyond the scope of this book. We will content ourselves with the discussion of some simple but non-trivial examples of analytic continuation in this sense (see Dwork [29] for more).

6.4.1 An example of Dwork

Consider the power series

$$F(x) := \frac{1}{\sqrt{1-4x}} = \sum_{n \geq 0} \binom{2n}{n} x^n \qquad (6.67)$$

(the factor of 4 is not essential but has the advantage of making the coefficients on the right-hand side be integers, see Ex. 19). Fix a prime $p > 2$. The series (6.67) defines F as an locally analytic function in the disk $D^- = \{x \mid |x|_p < 1\}$ and hence the same is true for $f(x) = F(x)/F(x^p)$. In fact, f has an analytic continuation to a larger domain. From [29], §3 we find that

$$F(x)F_s(x^p) \equiv F_{s+1}(x)F(x^p) \bmod p^{s+1}\mathbf{Z}_p[[x]], \qquad (6.68)$$

where

$$F_s(x) := \sum_{n=0}^{p^s-1} \binom{2n}{n} x^n \qquad (6.69)$$

is the truncation of the series for F to order $O(x^{p^s})$. Moreover, let $D_1 := \{x \mid |F_1(x)|_p = 1\}$. Then, uniformly in D_1, we have

$$f(x) = \lim_{s \to \infty} \frac{F_{s+1}(x)}{F_s(x^p)}. \qquad (6.70)$$

In particular, if we take $s = 0$ in (6.68) we find

$$f(x) \equiv F_1(x) \bmod p \qquad (6.71)$$

and since

$$\binom{\frac{1}{2}(p-1)}{n} \equiv \binom{-\frac{1}{2}}{n} \equiv \frac{1}{(-4)^n}\binom{2n}{n} \bmod p \qquad (6.72)$$

we also have

$$F_1(x) \equiv (1-4x)^{\frac{1}{2}(p-1)} \bmod p. \qquad (6.73)$$

Take now $\xi = \text{Teich}(a)$ for $a \in \mathbf{F}_p$ such that $1 - 4a \neq 0$. Since $\xi^p = \xi$ we may be tempted to believe that $f(\xi) = F(\xi)/F(\xi^p) = 1$ but this is in general not the case. The point is that though f extends analytically to a domain containing ξ, F itself does not. To compute $f(\xi)$ we note that $f^2(\xi) = (1-4\xi^p)/(1-4\xi) = 1$, since F^2 does extend to a domain containing ξ. Hence $f(\xi) = \pm 1$. To determine what is the correct sign it is enough to know its value mod p. Using (6.73) and Euler's criterion for the Legendre symbol we find finally that

$$f(\text{Teich}(a)) = \left(\frac{1-4a}{p}\right), \quad a \in \mathbf{F}_p, \quad 1 - 4a \neq 0. \qquad (6.74)$$

6.4.2 A generalization

We can repeat the calculations of the previous section in a different way and in a slightly more general case. Fix, as before, a prime $p > 2$, pick $h \in \mathbf{Z}_p[x]$ a polynomial with $h(0) = 1$ and let

$$F(x) := h(x)^{-\frac{1}{2}} = 1 + a_1 x + a_2 x^2 + \cdots \in \mathbf{Z}_p[[x]]. \qquad (6.75)$$

We now give the analytic continuation of $f(x) := F(x)/F(x^p)$ to a larger domain and verify the analogue of (6.74).

We claim that for all $s = 0, 1, \ldots$ we have the following approximation of f by rational functions

$$f(x) \equiv \frac{h(x)^{\frac{1}{2}(p^{s+1}-1)}}{h(x^p)^{\frac{1}{2}(p^s-1)}} \mod p^{s+1} \mathbf{Z}_p[[x]]. \qquad (6.76)$$

This congruence follows from the following

$$\left(\frac{F(x)^p}{F(x^p)} \right)^{p^s} \equiv 1 \mod p^{s+1} \mathbf{Z}_p[[x]], \qquad (6.77)$$

which is easy to verify by induction on s (see also [20], Ex. 6, p. 311).

If $\xi = \mathrm{Teich}(a)$ for $a \in \mathbf{F}_p$ with $h(a) \neq 0$ then the right-hand side of (6.74) for $x = \xi$ gives

$$h(\xi)^{\frac{1}{2} p^s(p-1)} = \left(h(\xi)^{\frac{1}{2}(p-1)} \right)^{p^s} \mod p^{s+1} \qquad (6.78)$$

and taking the limit as $s \to \infty$ we conclude that

$$f(\mathrm{Teich}(a)) = \left(\frac{h(a)}{p} \right), \quad a \in \mathbf{F}_p, \quad h(a) \neq 0. \qquad (6.79)$$

6.4.3 Dwork's exponential

A less trivial and very useful example of analytic continuation is Dwork's modified exponential function.

Choose $\pi \in \mathbf{C}_p$ such that $\pi^{p-1} = -p$ and define the *Dwork exponential* as

$$\Theta_p(x) = \exp(\pi(x - x^p)), \quad |x|_p < 1. \qquad (6.80)$$

Note that

$$\Theta_p(x) = \sum_{n \geq 0} c_n \pi^n x^n, \qquad (6.81)$$

with c_n as in (6.13). It follows from (6.21) that this power series converges in a disk containing $D := \{x \mid |x|_p \leq 1\}$ thereby extending the original domain of definition of Θ_p.

In particular, for example, $\Theta_p(1)$ makes sense; call this value ζ_p. As in §6.4.1 a rash judgment might lead one to believe that $\zeta_p = 1$ since, after all, $1 - 1^p = 0$ and $\exp(0) = 1$. However, this is not the case. The point is that $\Theta_p(x)$ and the composition, say T, of the functions $x \mapsto x - x^p$ and $y \mapsto \exp(\pi y)$ agree on the unit disk $D^- := \{x \mid |x|_p < 1\}$ but not necessarily beyond this domain. Even at points with $|x|_p = 1$ where $T(x)$ might be defined (for example at $x = 1$) it need not agree with the extension of T to a larger domain given by the power series for Θ_p.

In fact, ζ_p is a primitive p-th root of unity. To see this, raise it to the p-th power to get $\zeta_p^p = \Theta_p(1)^p$. Let $F(x) = \exp(\pi p x)$; it is given by a power series which converges on a disk containing D. Both $\Theta_p(x)^p$ and $F(x)/F(x^p)$ are therefore analytic on a disk containing D and agree on D^- hence they must agree on D as well. In particular, $\zeta_p^p = F(1)/F(1^p) = 1$. (Compare this discussion with that in §6.4.1.)

On the other hand, by (6.81) we have

$$\zeta_p = \Theta_p(1) \equiv 1 + \pi \bmod \pi^2 \qquad (6.82)$$

(note that $c_n = 1/n!$ for $n < p$ so $c_1 = 1$ for all p); hence, in particular, $\zeta_p \neq 1$. The same argument shows that in general $\Theta(\mathrm{Teich}(a))$ is a primitive p-th root of unity for any $a \in \mathbf{F}_p^\times$. We have that

$$\Theta_p(\mathrm{Teich}(a)) \equiv 1 + a\pi \bmod \pi^2. \qquad (6.83)$$

Hence, we have established a bijection between the primitive p-th roots of unity in $\mathbf{C}_p$ and the roots of the polynomial $x^{p-1} + p = 0$; namely, $\Theta_p(\zeta) \leftrightarrow \zeta \pi$ with ζ running over primitive $(p-1)$-st roots of unity in $\mathbf{Z}_p^\times$. We also deduce that

$$\Theta_p(1)^a = \Theta_p(\mathrm{Teich}(a)), \quad a \in \mathbf{Z}/p\mathbf{Z} \qquad (6.84)$$

(since $\Theta_p(1)$ is a p-th root of unity the left-hand side is unambiguously defined). In other words, we obtain an additive character

$$\psi : \mathbf{F}_p \longrightarrow \mathbf{C}_p^\times$$
$$a \mapsto \Theta_p(\mathrm{Teich}(a)) = \zeta_p^a \qquad (6.85)$$

(additive here refers to the fact that we consider the source $\mathbf{F}_p$ as a group with addition; i.e., we have $\psi(a+b) = \psi(a)\psi(b)$). This character naturally depends on the choice of π, or, equivalently, of ζ_p. We will call any such choice a *Dwork character*. It is the p-adic analogue of the complex additive character

$$\mathbf{F}_p \longrightarrow \mathbf{C}^\times$$

$$a \mapsto \exp(2\pi i a/p)$$

(which of course also depends of the choice $\exp(2\pi i/p)$ of primitive p-th root of 1).

Remark A completely analogous analysis holds for an arbitrary finite field $\mathbf{F}_q$ with $q = p^f$ instead of $\mathbf{F}_p$. We only need to replace Θ_p by

$$\Theta_q(x) := \exp(\pi(x - x^q)). \tag{6.86}$$

6.5 Gauss sums and the Gross–Koblitz formula

To simplify the exposition we first consider the case of the finite field $\mathbf{F}_p$; in the next section, we discuss the general finite field.

6.5.1 The case of $\mathbf{F}_p$

A *Gauss sum* on $\mathbf{F}_p$ is a sum of the form

$$g(\chi, \psi) := -\sum_{x \in \mathbf{F}_p^\times} \chi(x)\psi(x) \tag{6.87}$$

where χ is a multiplicative character (a character of $\mathbf{F}_p^\times$) and ψ is a nontrivial additive character (a character of $\mathbf{F}_p$). Usually, we consider χ and ψ as taking values in $\mathbf{C}^\times$ and hence $g(\chi, \psi) \in \mathbf{C}$ but, as we will see below, it is also useful to consider the case where they have values in $\mathbf{C}_p$ and hence also $g(\chi, \psi) \in \mathbf{C}_p$. (The minus sign is convenient and not entirely standard; it makes (6.90) below true.)

Gauss sums are pretty ubiquitous. In a sense, they are the mod p analogues of the complex gamma function $\Gamma(s)$. Indeed, we have

$$\Gamma(s) = \int_0^\infty e^{-t} t^s \frac{dt}{t}, \quad \Re(s) > 0 \tag{6.88}$$

and we may view this as the integral of the product of an additive character and a multiplicative character with respect to the invariant measure dt/t on $\mathbf{R}_{>0}^\times$.

In fact this analogy carries on further. We will now describe the Gross–Koblitz formula which expresses $g(\chi, \psi)$, now viewed as an element of $\mathbf{C}_p$, in terms of the p-adic gamma function.

We begin with some elementary facts about Gauss sums. The dependence of $g(\chi,\psi)$ on ψ is fairly mild. Any other additive character is of the form ψ^a for some integer a prime to p. Then as it is easily seen

$$g(\chi,\psi^a) = \chi(a)^{-1} g(\chi,\psi). \tag{6.89}$$

For the trivial character $\chi = 1$ we have

$$g(1,\psi) = 1, \tag{6.90}$$

and if χ is not trivial

$$|g(\chi,\psi)|^2 = p, \quad \chi \neq 1. \tag{6.91}$$

Let us fix once and for all $\pi \in \mathbf{C}_p$ a root of $x^{p-1} + p = 0$ and ψ the associated Dwork character as in (6.85). Then in $\mathbf{C}_p$, and dropping ψ from the notation, a Gauss sum has the form

$$g_a := -\sum_{x \in \mathbf{F}_p^\times} \mathrm{Teich}(x)^{-a} \psi(x), \quad a \in \mathbf{Z}, \tag{6.92}$$

or, equivalently,

$$g_a = -\sum_{\zeta^{p-1}=1} \zeta^{-a} \Theta_p(\zeta). \tag{6.93}$$

The properties of Gauss sums (6.90) and (6.91) become

$$g_0 = 1, \quad g_a g_{-a} = (-1)^a p, \quad 0 < a < p-1. \tag{6.94}$$

Finally, we may now state the Gross–Koblitz formula (over $\mathbf{F}_p$, see [43], [78])

Theorem 6.4 *With the above notation we have*

$$g_a = \pi^a \Gamma_p\left(\frac{a}{p-1}\right), \quad 0 \leq a < p-1. \tag{6.95}$$

We leave it as an exercise (Ex. 20) to verify that this formula is consistent with (6.94). It is now clear that $\Gamma_p(\frac{a}{p-1})$ is algebraic (the nature of $\Gamma_p(r)$ for a general $r \in \mathbf{Q}$ is unknown).

As a corollary we find the following result of Stickelberger

Corollary 6.5

$$v_p(g_a) = \frac{a}{p-1}, \quad 0 \leq a < p-1. \tag{6.96}$$

The proof is immediate since $\Gamma_p\left(\frac{a}{p-1}\right) \in \mathbf{Z}_p^\times$.

6.5.2 An example

We give now a simple example of using the Gross–Koblitz formula for computational purposes. Though certainly not the most efficient way to solve the equation $4p = u^2 + 27v^2$ in integers u, v (see for example fermat in Chapter 1 for a similar question) it suggests the possible use of p-adic methods for the computation of interesting integers (see also [17]).

Fix a prime $p \equiv 1 \bmod 3$ and a non-trivial additive character $\psi : F_p \to C^\times$. It was known already to Gauss [26], (4.24) that if χ is a multiplicative character modulo p of order 3 then

$$g(\chi, \psi)^3 = p\alpha, \qquad \alpha \in Z[\zeta_3], \qquad \alpha\bar{\alpha} = p. \qquad (6.97)$$

Here $\zeta_3 \in C$ is a fixed cubic root of 1. Note that by (6.89) α is independent of the choice of ψ.

We can compute α as a p-adic integer using Γ_p and the Gross–Koblitz formula. Indeed, taking χ to be the n-th power, with $n := \frac{1}{3}(p-1)$, of the Teichmüller character we have

$$\alpha = -\Gamma_p\left(\tfrac{1}{3}\right)^3. \qquad (6.98)$$

It is known that, in fact, $\alpha = \frac{1}{2}(u + \sqrt{-27}v)$ with $u, v \in Z$, $u \equiv -1 \bmod 3$, satisfying $4p = u^2 + 27v^2$. The following GP routines compute one such pair (u, v) (there are two pairs, the other is $(u, -v)$) for a given p.

```
gsumc(p) =
{
  local(g);
  g=-gammap(1/3+O(p))^3;
  [centerliftp(g+p/g),centerliftp((g-p/g)/sqrt(-27+O(p)))]
}
```

Here we used the ad hoc function

```
centerliftp(x) =
{
  local(a,b);
  a=truncate(x);
  b=truncate(-x);
  if(a < b,a,-b)
}
```

to recover an integer from its p-adic expansion. The built-in GP function truncate(x) returns a positive integer congruent to x to the given precision; if x is the p-adic expansion of a negative integer we hence want -truncate(-x). If we had the expansion of a rational number instead of an integer we could easily adapt the function recognizemod of §1.2.2 to this situation.

For example, we find

```
? forprime(p=3,100,if(p%3==1, print(p,"\t", gsumc(p))))
7         [-1, -1]
13        [5, 1]
19        [-7, -1]
31        [-4, -2]
37        [11, 1]
43        [8, 2]
61        [-1, 3]
67        [5, 3]
73        [-7, 3]
79        [17, 1]
97        [-19, -1]
```

6.6 Exercises

1. (Hard) Let $f(n) := \sum_{k=0}^{n-1} \binom{2k}{k}/\binom{2n}{n}$ for $n \in \mathbb{N}$. Show f extends to a continuous function from $\mathbb{Z}_3$ to itself.
2. For $a \in \mathbb{Z}_p$ let $f(a)$ be the unique root in $\mathbb{Z}_p$ of the equation $x^p + x = a$. Find a closed form for the coefficients in the power series expansion of f about $a = 0$. Prove that this power series converges in the disk $|a|_p < 1$.
3. Prove that $v_p(n!) = \sum_{k=1}^{\infty} \left\lfloor \frac{n}{p^k} \right\rfloor$ and deduce that $v_p(n!) \leq n/(p-1)$.
4. Prove that for all $n \in \mathbb{N}$
$$a_n := \frac{(30n)!n!}{(15n)!(10n)!(6n)!} \in \mathbb{Z}.$$
5. Show that with n defined by (6.22), $n - \lfloor n/p \rfloor - m$ equals 0 or 1.
6. For $n \in \mathbb{N}$ let $f_p(n) := n!/p^{v_p(n!)}$. Write a routine that computes $f_p(n)$ in terms of Γ_p. Compute $f_7(10^{10})$ to order $O(7^{20})$.
7. Prove that $\Gamma_p(\frac{1}{2})^2 = -\left(\frac{-4}{p}\right)$ for all $p > 2$. Study the statistics of the sign $\Gamma_p(\frac{1}{2})$ as p varies.
8. Prove that c_n of (6.13) is the number of solutions of $\sigma^p = 1$ in the symmetric group S_n.
9. (K. Belabas) Show that the following (divide-and-conquer) routine computes $\lfloor \log_p(x) \rfloor$.

```
flogp(x,p) = aux(x,p)[1]

aux(x,p) =
{
    if (x < p, return([0,x]));
```

```
        v = aux(x, p^2);
        if (v[2] >= p, [2*v[1]+1, v[2]\p]
                     , [2*v[1]  , v[2]]);
}
```
and
```
clogp1(x,p) = local(k); k=flogp2(x,p); if (p^k < x, k+1, k)
```
computes $\lceil \log_p(x) \rceil$. Compare its running time with that of `clogp` given in §6.3.

10. Extend the test of Diamond's formula (6.40) to the case when p is inert in the real quadratic field F.
11. Give an elementary proof of Wolstenholme's theorem (6.53).
12. Prove (6.47).
13. Write a routine to find a p-adic approximation of the coefficients b_n of (6.43) for $p = 3$ and confirm those given in the text.
14. Prove that $H_n \notin \mathbf{Z}$ for $n > 1$.
15. Write a routine that computes sums of the form $\sum_{k=1}^{n} k^{-2}$ p-adically.
16. Prove (6.65) and find an expression for $\lim_{s \to \infty} A(p^s)$.
17. Prove Glaisher's congruence
$$\binom{np+p-1}{p-1} \equiv 1 \bmod p^3, \quad n \in \mathbf{N}, \quad p > 5.$$
18. The *index of irregularity* r_p of a prime p is the number of Bernoulli numbers $B_2, B_4, \ldots, B_{p-3}$ that are divisible by p. A prime p is *regular* if $r_p = 0$ and *irregular* otherwise. Using the identity
$$B_n(x) = \sum_{k=0}^{n} \binom{n}{k} B_k x^{n-k} \qquad (6.99)$$
write a function based on `bernoullip` that computes $B_2, B_4, \ldots, B_{p-3}$ modulo p. Use this to write a function that computes r_p and find all irregular primes $37, 59, 67, 101, \ldots$ up to, say, 1000.
19. If $f(x) = 1 + \sum_{n \geq 1} a_n x^n$ is the power series expansion of an algebraic function with $a_n \in \mathbf{Q}$ then a theorem of Eisenstein (see [20], Theorem 5.1) guarantees that for some $N \in \mathbf{N}$ the power series expansion of $f(Nx)$ will have integer coefficients. Find the smallest such N for $f(x) = (1-x)^{1/p}$ with p prime.
20. Verify that the Gross–Koblitz formula (6.95) is consistent with (6.94), using (6.10).

7 Polynomials

7.1 Mahler's measure

For a non-zero polynomial $P = a\prod_\alpha (x - \alpha) \in \mathbf{C}[x]$ we define its *Mahler measure* as

$$M(P) = |a| \prod_{|\alpha|>1} |\alpha| = \exp\left(\int_0^1 \log |P(e^{2\pi i\theta})|\, d\theta\right). \tag{7.1}$$

This quantity measures in a certain sense the size of P; other measures are for example,

$$L(P) = \sum_{i=0}^n |a_i|, \quad H(P) = \max_{i=0,\ldots,n} |a_i|, \quad ||P||_\infty = \max_{|x|=1} |P(x)| \tag{7.2}$$

where $P(x) = a_n x^n + \cdots + a_0$. An advantage of M, however, is that it is visibly multiplicative

$$M(P_1 P_2) = M(P_1) M(P_2). \tag{7.3}$$

The basic inequalities relating these measures are

$$M(P) \leq ||P||_\infty \leq L(P) \leq 2^n M(P). \tag{7.4}$$

The measure was introduced by Mahler [61] to give a simple proof of an inequality due to Gelfond.

For $P \in \mathbf{Z}[x]$ we have $M(P) \geq 1$ with equality if and only if P is monic and all of its non-zero roots are in the unit circle. By a theorem of Kronecker therefore $M(P) = 1$ if and only if P is a product of a power of x and a cyclotomic polynomial.

If $P(x) = ax + b$ then

$$M(P) = \max\{|a|, |b|\}. \tag{7.5}$$

For a, b relatively prime integers the right-hand side is called the *height* of the rational number $\alpha = -b/a$ (the unique root of P). What $M(P)$ measures has little to do with the size of α in the usual sense but rather it has to do with its *complexity*; note that $\log M(P)$ is roughly proportional to how long it would take to communicate α (with a and b as a string of decimal digits) to somebody else. In general for $P \in \mathbf{Z}[x]$ irreducible with relatively prime coefficients, $M(P)^{1/n}$ is the height of any of its roots.

A famous question of Lehmer [58] asks whether there is a minimum for $M(P)$ as P runs through all non-zero polynomials with integer coefficients with $M(P) > 1$ (check Mossinghoff's Lehmer's question website [67] for more information about this problem). To date the integer polynomial with the smallest value of $M(P) > 1$ is the degree ten polynomial found by Lehmer himself in his original paper, namely,

$$M(x^{10} + x^9 - x^7 - x^6 - x^5 - x^4 - x^3 + x + 1) = 1.1762801\ldots \quad (7.6)$$

7.1.1 Simple search

It is not hard to prove that

$$|a_i| \leq \binom{n}{i} M(P), \quad P(x) = a_n x^n + \cdots + a_0. \quad (7.7)$$

In particular, the set of $P \in \mathbf{Z}[x]$ of degree n and $M(P) \leq M$ for a fixed $M \geq 1$ is finite. Here is a simple GP routine that will run through this set of polynomials for a given n.

```
mahlerlist(n, M) =
{
  forvec(v = vector(n,k, floor(M*binomial(n,k)) * [-1,1]),
    f = x^n + Pol(v);
    m = polmahler(f);
      if (m > 1 && m < M && polisirreducible(f),
        print(f, "\t", m)));
}
```

It uses the following function to calculate Mahler's measure straight from the definition

```
polmahler(f) =
{
  local(rv);
  rv=abs(polroots(f));

  polcoeff(f,poldegree(f))*
  prod(k=1,length(rv),if(rv[k]>1,rv[k],1))
}
```

We should warn the reader that the collection of polynomials satisfying (7.7) gets very big quickly as n increases. For example, if we want all polynomials of degree n with $M < 2$ then we may reduce to searching polynomials $P(x) = a_n x^n + \cdots + a_0$ with $a_n = 1, a_0 = \pm 1$ and $a_i < 2\binom{n}{i}$ for $i = 1, 2, \ldots, n-1$. The total number of such polynomials is $2 \prod_{i=1}^{n-1}(2\binom{n}{i}) - 1)$ with values

2, 6, 50, 1078, 58482, 7937358, 2705095458, 2331302613750, ...

for $n = 1, 2, \ldots$.

We can cut down this set by exploiting the fact that $M(P)$ does not change if we replace $P(x)$ by $P(-x)$ or $P(1/x)x^n$ but it will still be pretty big. Nevertheless, with $n = 5$ and $M \leq 1.5$ mahlerlist took under 20 minutes to run. Here is a short list sorted by value of M (after removing cyclotomic factors which do not change the value of M).

P	M
$x^3 - x + 1$	1.3247179572447460259609088 53...
$x^5 - x^4 + x^3 - x + 1$	1.3497161046696958652982404 39...
$x^5 - x^2 + 1$	1.3641995455827723418260742 24...
$x^4 + x - 1$	1.3802775690976141156733016 90...
$x^5 - x + 1$	1.4098717208302613338696888 49...
$x^5 + x^4 - x^3 - x^2 + 1$	1.4432687912703731076281276 06...
$x^3 + x - 1$	1.4655712318767680266567312 24...

Performing the search for polynomials with small $M > 1$ in this way makes Lehmer's discovery of the polynomial (7.6) seem to be nothing short of miraculous. Though there is no hint in the original paper [58] one might make a guess of what other search spaces Lehmer could have used (say, polynomials with coefficients $-1, 0, 1$ or even shifted cyclotomics as in Ex. 4) but this does not make the discovery less impressive, particularly, if we consider the date (1930s).

7.1.2 Refining the search

We sketch here how can we speed up the search of our previous section following Boyd [13] (see also Mossinghoff's talk [66]). Define a polynomial P to be *reciprocal* if $P^* = \pm P$, where $P^*(x) := x^n P(1/x)$ and n is the degree of P. By a theorem of Smyth [87] if P is not reciprocal (and not x^n for some n) then

$$M(P) \geq M(x^3 - x + 1) = 1.3247\ldots \tag{7.8}$$

We can therefore restrict our search to reciprocal polynomials. If $P^* = -P$ then $P(1) = 0$ and if $P^* = P$ but P is of odd degree then $P(-1) = 0$; hence, we may, in fact, restrict to polynomials of the form

$$P(x) = x^{2m} + 1 + \sum_{k=1}^{m-1} a_k(x^k + x^{2m-k}). \tag{7.9}$$

Also, we should exploit the fact that $M(P(x^k)) = M(P(x))$ for any $k \in \mathbb{N}$ and $M(PQ) = M(P)$ if Q is cyclotomic.

A more substantial refinement, due to Boyd [13], arises from the bound

$$|a_k| \leq \binom{n}{k} + \binom{n-2}{k-1}(M^2 + M^{-2} - 2), \quad \text{if } M(P) \leq M, \tag{7.10}$$

an improvement on (7.7), combined with the following idea.

For any $r \in \mathbb{N}$ let

$$G_r P(x) := \text{Res}_t(P(t), x - t^r). \tag{7.11}$$

In GP this would be

```
polgraeffer(f,r)=polresultant(subst(f,x,t),x-t^r,t)
```

Here $\text{Res}_t(f, g)$ denotes the resultant of f, g in the variable t. It is not hard to verify that the defining property of $G_r P$, for P monic, is that it is a monic polynomial of the same degree, whose roots (counting with multiplicity) are the r-th power of those of P. In general, the leading coefficient of $G_r(P)$ is the r-th power of leading coefficient of P.

The case $r = 2$, where the roots of $GP := G_2 P$ are the square of those of P, is used in the *Graeffe root-squaring* method for finding roots of polynomials. For us the point is that $GP(x)$ can be computed very fast (much faster than computing $M(P)$) and satisfies

$$M(GP) = M(P)^2, \quad \text{and} \quad M(P) = \lim_{k \to \infty} \sqrt[2^k]{L(G^k P)}. \tag{7.12}$$

Hence to screen polynomials P for values of $M(P) < M$ for a given M we may test (7.10) not only for P but also for $G^k P$ for $k = 2, \ldots, m$ for some fixed m (with M replaced by M^{2^k}).

We may compute $GP(x)$ directly without appealing to resultants as follows (see Ex. 1)

$$GP(x) = (-1)^n(a(x)^2 - xb(x)^2), \quad \text{if } P(x) = a(x^2) + xb(x^2) \text{ of degree } n. \tag{7.13}$$

7.1 Mahler's measure

The following is a straightforward implementation of this in GP.

```
polgraeffe(f) =
{
  local(v,d,hd);
  d=poldegree(f);
  hd=d\2;

  v=[sum(k=0,hd,polcoeff(f,2*k)*x^k),
  sum(k=0,hd,polcoeff(f,2*k+1)*x^k)];

  f=v[1]^2-x*v[2]^2;

  if(d%2,-f,f)
}
```

Here is the complete list of all conic irreducible polynomials P with $1 < M(P) < 1.3$ and $\deg(P) \le 12$, as determined by Boyd [13] and extracted from [67] (see that page for a similar list for higher degree polynomials).

deg	M	O	Coefficients
8	1.280638156268	1	1 0 0 1 -1
10	1.176280818260	1	1 1 0 -1 -1 -1
10	1.216391661138	1	1 0 0 0 -1 1
10	1.230391434407	1	1 0 0 1 0 1
10	1.261230961137	1	1 0 -1 0 0 1
10	1.267233859440	2	1 0 1 1 0 1
10	1.283582360621	2	1 1 0 0 0 -1
10	1.293485953125	1	1 0 -1 1 0 -1
12	1.227785558695	2	1 1 1 0 -1 -1 -1
12	1.240726423653	1	1 1 1 1 0 0 -1
12	1.251046617204	2	1 1 0 0 0 -1 -1
12	1.264393854743	2	1 0 1 0 0 1 -1
12	1.272818365083	2	1 0 1 1 1 2 1

O denotes the number of roots of the polynomial outside the unit circle. Only half of the coefficients appear in the table (to be read left to right) since the polynomials are reciprocal. To be safe, the first polynomial is

```
? f=1+x^3-x^4+x^5+x^8;
? polmahler(f)

 1.2806381562677575967019025333

? abs(polroots(f))
```

```
[1.2806381562677575967019025330,  0.7808606944168846216260608221,
1.0000000000000000000000000000,  1.0000000000000000000000000000,
1.0000000000000000000000000000,  1.0000000000000000000000000000,
1.0000000000000000000000000000,  1.0000000000000000000000000000]~
```

7.1.3 Counting roots on the unit circle

The last calculation raises the question of how to count the number of roots a real polynomial has on the unit circle without computing the roots themselves. This is analogous to counting the number of real roots of a real polynomial, something that can be done using Sturm's method. In fact, we can actually reduce our problem to finding the real roots of an associated polynomial and hence solve it this way. (There are more direct methods but we have not traced their literature.)

We may assume without loss of generality (see Ex. 2) that P is reciprocal and $P(1) \neq 0$. Parameterize the unit circle $|x| = 1$ (except for $x = 1$) as follows

$$x = \frac{t^2 - 1}{t^2 + 1} + \frac{2it}{t^2 + 1}, \quad t \in \mathbf{R}. \tag{7.14}$$

Make this substitution (7.14) in $P(x)$ to obtain $R(t)$, a rational function of t. Let Q be the numerator of the real part of R. Now apply Sturm's method to Q. (Since P is reciprocal the numerators of $\Im(R(t))$ and $\Re(R(t))$ are proportional and hence have the same zeros.)

Here is what we get for the polynomial at the end of §7.1.2.

```
? Q=numerator(real(subst(f,x,(t^2-1)/(t^2+1)+2*I*t/(t^2+1))));
? polsturm(Q)

4
```

7.2 Applications of the Graeffe map

We discuss several applications of the general Graeffe map G_r of polynomials defined in (7.11). For the first two other more straightforward methods are also possible (see [10]). We simply show how to solve the problems using G_r.

7.2.1 Detecting cyclotomic polynomials

We can use the operation G_r with $r > 1$ on polynomials to detect whether a polynomial is cyclotomic and also to find the largest cyclotomic factor of a given polynomial. (See also Ex. 5.20) The basic fact, which is easily shown,

is that
$$G_r\Phi_m = \Phi_{m'}^e, \quad m' := m/\gcd(m,r), \quad e = \phi(m)/\phi(m'), \quad (7.15)$$
where Φ_m is the m-th cyclotomic polynomial. Since we also have that
$$G_r(PQ) = (G_rP)(G_rQ) \quad (7.16)$$
(see Ex. 1) it follows that if $P = x^{m_0}\prod_i \Phi_{m_i}$ and r is coprime to $m_1 m_2 \cdots$ then
$$G_r P = P. \quad (7.17)$$
Conversely, if P is non-zero and $G_r P = P$ for some $r > 1$ then P is of the form $P = x^{m_0}\prod_i \Phi_{m_i}$ with r coprime to $m_1 m_2 \cdots$.

This leads to the following algorithm to determine whether $P \in \mathbb{Z}[x]$ is a product of cyclotomic polynomials and powers of x.

```
poliscyclo(f)=
{
  local(f1,k,n);
  f=f/x^valuation(f,x);
  n=poldegree(f);

  if(abs(polcoeff(f,0)*polcoeff(f,n)) > 1, 0,

  k=1; n=shift(n,2);

  until(f1 == f || k > n,

    f1=f;
    f=polgraeffe(f1);
    k=shift(k,1));

    if(k > n,0,1))
}
```

The worst case to analyze with this method is for $P = \Phi_{2^r}$ of degree $n = 2^{r-1}$. The first time that (7.17) is satisfied among $G^j P$ with $j = 0, 1, \ldots$ is for $G(G^r P) = G^r P$. Hence we must allow k, which equals 2^s with s the number of steps in the loop, to be as large as 2^{r+1} or $4n$. (This explains the line n=shift(n,2) in the above code.) We go through the main loop in the algorithm at most $\log_2(n) + 2$ steps. Note that a product of cyclotomic polynomials has integer coefficients, is monic and has constant term ± 1. The line of code

```
if(abs(polcoeff(f,0)*polcoeff(f,n)) > 1,
```

checks whether this is satisfied before performing the actual test in terms of G.

7.2.2 Detecting cyclotomic factors

In the same vein, we can write an algorithm that will compute the largest product of cyclotomic polynomials that divides a given polynomial $P \in \mathbf{Z}[x]$. Indeed, $\gcd(P, G_r P)$ is the largest product of cyclotomic polynomials $\Phi_{m_1} \Phi_{m_2} \cdots$ dividing P with $m_1 m_2 \cdots$ coprime to r. Hence we iteratively make

$$P \mapsto P/\gcd(P, G_r P)$$

with $r = 2, 3, 5, \ldots$ running over the prime numbers. Stopping this process at $r = R$ will leave a polynomial P_R which if divisible by Φ_m with $m > 1$ must have m divisible by $r = 2, 3, 5, \ldots, R$ and, in particular, $\deg(P_R) \geq \prod_{r=2}^{R}(r-1)$. Hence, if $\prod_{r=2}^{R}(r-1) > \deg(P_R)$ then P_R has no cyclotomic factors. The following is a GP implementation of this idea.

```
polcyclofactor(f)=
{
  local(g,h,k,p,l,n);

  g=1; k=1; l=1; f=f/x^valuation(f,x);

  until(l > n ,

    p=prime(k);l*=p-1;k++;
    h=gcd(f,polgraeffer(f,p));
    if(poldegree(h),f/=h;g*=h);
    n=poldegree(f));

  g
}
```

7.2.3 Wedge product polynomial

Let $f = \prod_{\nu=1}^{n}(x - \alpha_\nu) \in \mathbf{C}[x]$ be a polynomial. As a final application of G_r we consider the problem of computing the polynomial f_k whose roots are the products $\alpha_{\nu_1} \alpha_{\nu_2} \cdots \alpha_{\nu_k}$ of k distinct roots of f. We naturally want to do this without actually computing the roots of f. (The reason for the name polwedge is that if f is the characteristic polynomial of a diagonalizable γ acting on an n-dimensional C-vector space V then f_k is the characteristic polynomial of γ acting on $\bigwedge^k V$.)

As we pointed out in §7.1.2 we have

$$G_r f(x) = \prod_{\nu=1}^{n}(x - \alpha_\nu^r). \tag{7.18}$$

7.2 Applications of the Graeffe map

The k-th coefficient of $G_r f$ is then

$$(-1)^{n-k} \sum_{v_1 < v_2 < \cdots < v_k} \alpha_{v_1}^r \alpha_{v_2}^r \cdots \alpha_{v_k}^r.$$

This is the r-th power sum of the roots of f_k. Hence since f_k obviously has degree $\binom{n}{k}$ we can reconstruct it using Newton's formulas if we know $G_r f$ for $r = 1, 2, \ldots, \binom{n}{k}$. In fact, we can compute all f_j with $j \leq k$ at the same time. Here is a GP implementation.

```
polwedge(f,l) =
{
  local(n,hn,v,ff,m,fv,a0);

  n = poldegree(f); a0=polcoeff(f,0);
  m=n; a0 = if(n%2,-a0,a0); if(l, ,l=n);   hn=min(l,n\2);
  v = concat(f,vector(binomial(n,hn)-1,r,polgraeffer(f,r+1)));

  fv=vector(l+1);
  fv[1]=x-1;fv[2]=f;
  if(l >= n, fv[n+1]=x-a0);

  for(k=2,min(l,n-1),

  m*=(n-k+1)/k;

  ff=exp((-1)^(k-1)*sum(r=1,m,
       polcoeff(v[r],n-k)*x^r/r,O(x^(m+1)))));

  ff=truncate(ff);
  ff=ff/polcoeff(ff,0);
  ff=subst(ff,x,1/x)*x^m;

  fv[k+1]=ff);

  fv
}
```

In the program we use the Newton formulas expressed as the truncation of the following identity

$$\prod_{v=1}^{m}(1 - \gamma_v T) = \exp\left(-\sum_{r=1}^{\infty}\frac{N_r}{r}T^r\right), \quad \text{if } N_r = \sum_{v=1}^{m}\gamma_v^r. \tag{7.19}$$

7.2.4 Interlacing roots of unity

We would like to answer the following question. Find all pairs of finite sets $S, T \subseteq \overline{\mathbf{Q}} \subseteq \mathbf{C}$ of roots of unity which have the same cardinality

#S = #T = r, are Gal($\overline{\mathbf{Q}}/\mathbf{Q}$)-stable (i.e., for all $s \in S$ and $\sigma \in$ Gal($\overline{\mathbf{Q}}/\mathbf{Q}$) we have $s^\sigma \in S$ and similarly with T) and such that the elements of S and T interlace in the unit circle (i.e., as we traverse the unit circle the elements of S and T alternate ... s, t, s', t' ...; in particular the sets must be disjoint).

As a simple example with $r = 2$ we have $S = \{1, -1\}, T = \{i, -i\}$. How can we find all examples for a given size r? In §5.4.2 we described a function that computes all cyclotomic polynomials of a given degree r. We can modify this function so that it only gives cyclotomic polynomials of degree r with no repeated factors (i.e., with no double roots). The set of roots of such a polynomial precisely corresponds to a Gal($\overline{\mathbf{Q}}/\mathbf{Q}$)-stable set of roots of unity of size r and conversely. To find all solutions to our question we could run through all possible pairs of such cyclotomic polynomials keeping the ones whose roots are interlaced.

How do we check whether the roots of two cyclotomic polynomials p, q of the same degree and with no repeated factors are interlaced? Rather than working with the roots of unity themselves we can consider their *exponents*: the sets A, B of rational numbers in the interval $[0, 1)$ such that

$$p = \prod_{a \in A}(x - e^{2\pi i a}), \quad q = \prod_{b \in B}(x - e^{2\pi i b}).$$

Then the roots of p and q are interlaced in the unit circle if and only if the sets A and B are interlaced in $[0, 1)$.

Here is a GP routine that checks interlacing in $[0, 1)$ of two sets of same cardinality. Given the input vectors A, B containing the ordered elements of the respective (multi) sets we form a vector w of pairs (x, y) where $x \in A \cup B$ and $y = 0$ if $x \in A$ and $y = 1$ if $y \in B$. We sort w according to its first entries x and then check if all consecutive pairs in the sorted vector are different. (We do not require the input sets to be disjoint so we also check that.)

```
interlacing(A,B)=
{
  local(w,k,s, r = length(A));
  if(r != length(B), error("Sets must be of the same size"));
  w=concat(vector(r,k,[A[k],0]), vector(r,k,[B[k],1]));
  w=vecsort(w,1);
  for (k = 2, length(w),
    if (w[k][1]==w[k-1][1] ||
        w[k][2]==w[k-1][2], return(0)));
  1;
}
```

In order to use this routine we need to find the exponents of a cyclotomic polynomial P. Assume first that we already know the expression P as $\prod_{v_i} \Phi_{v_i}$ with v_i not necessarily distinct. It is easy since the exponents of Φ_v are

7.2 Applications of the Graeffe map

simply all the rational numbers of the form k/v with $0 \le k < v$. Here is a GP routine that computes the exponents given the ordered vector of v_i's. (To make the routine more general we do not assume the polynomial is squarefree; the input is allowed to have repetitions.)

```
polcycloexp(v) =
{
  local(u,w,m);
  w=[];m=0;
  for(k=1,length(v),
    if(v[k] > m,
      m=v[k];
      u=[];
    if(v[k]==1,u=[0],
      for(n=1,v[k]-1,
        if(gcd(n,v[k])==1,u=concat(u,n/v[k])))));
  w=concat(w,u));
  w
}
```

For example

```
? polcycloexp([1,2,3,3,10])

[0, 1/2, 1/3, 2/3, 1/3, 2/3, 1/10, 3/10, 7/10, 9/10]
```

It is not hard to modify the function `polcyclolist` of §5.4.2 to give all squarefree cyclotomic polynomials of a given degree in the format we need to input into `polcycloexp` (see Ex. 6). Assume we have such a function, say `polcyclolist1`, already. Then we can finally find pairs of interlacing cyclotomic polynomials.

```
? polv=polcyclolist1(4)

[[1, 2, 3], [1, 2, 4], [1, 2, 6], [3, 4], [3, 6],
 [4, 6], [5], [8], [10], [12]]

? for(j=1,length(polv)-1,
    for(k=j+1,length(polv),
      if(interlacing(polcycloexp(polv[j]),
                     polcycloexp(polv[k])),
        print(polv[j],"\t",polv[k]))))

[1, 2, 3]    [5]
[1, 2, 3]    [8]
[1, 2, 3]    [12]
[1, 2, 4]    [3, 6]
[1, 2, 4]    [5]
```

[1, 2, 4]   [8]
[1, 2, 4]   [10]
[1, 2, 4]   [12]
[1, 2, 6]   [8]
[1, 2, 6]   [10]
[1, 2, 6]   [12]

7.3 Exercises

1. For $n \in \mathbf{N}$ let $K_n = \mathbf{C}(x^n)$, with x an indeterminate. Prove that
$$G_n P(x^n) = \pm N_{K_1/K_n}(P),$$
where $\mathbf{N}$ denotes the norm map. Verify that (7.13) and (7.16) are true.

2. Prove the assertion in §7.1.3 that we may assume without loss of generality that P is reciprocal and $P(1) \neq 0$.

3. Given a polynomial $P(x) = \prod_\alpha (x - \alpha) \in \mathbf{Z}[x]$ let
$$\Delta_n(P) := \prod_\alpha (\alpha^n - 1).$$
Prove that the $\Delta_n(P)$: (i) are integers; (ii) if P has no root on the unit circle, grow like $M(P)^n$; and (iii) satisfy a linear recurrence with constant coefficients.
Compute the linear recurrence satisfied by $\Delta_n(P)$ where $P = x^8 - x^5 - x^4 - x^3 + 1$.

4. Search for polynomials P with small $M(P) > 1$ among polynomials of the form $P(x) = Q(x) \pm x^m$ where Q is cyclotomic of degree m.

5. Graph and analyze the sequence $M(x^n + x + 1)$ for $n = 2, 3, \ldots$.

6. Write a GP function (named already `polcyclolist1` in the text) that will output all squarefree cyclotomic polynomials of a given degree (in the format $v_1 \leq v_2 \leq \cdots$ standing for $\Phi_{v_1} \Phi_{v_2} \cdots$).

7. Find the first n for which the cyclotomic polynomial Φ_n does not have coefficient $0, \pm 1$.

8. Find all interlacing pairs of cyclotomic polynomials of degree at most 8.

9. Write a GP script that converts a series $1 + a_1 x + a_2 + x^2 + \cdots$ with $a_i \in \mathbf{Z}$ to an infinite product $\prod_n (1 - x^n)^{-c_n}$ (to within the given precision). We can use this, for example, to write a cyclotomic polynomial as a product or Φ_n's.

8 Remarks on selected exercises

Ex. 1.5 See [42], §3.2 for a nice discussion of this and related matters.

Ex. 1.6 This is less trivial than it first appears. What exactly *is* $\sqrt[16]{-4}$? The question is not really well posed since the polynomial $x^{16}+4$ factors as

```
? factor(x^16+4)

[x^8 - 2*x^4 + 2 1]
[x^8 + 2*x^4 + 2 1]
```

This example is due to B. de Smit.

Ex. 1.13 The connection is in the general spirit of reciprocity as in §3.2.2 but goes beyond the scope of this book. See for example [26], (4.24) and chapter III, §14 C.

Ex. 1.26 Here is a simple GP script that will plot the ratio of $R_k(n)$ and the volume of the corresponding sphere for fixed k and $n = 1$ through a given bound

```
ssasymptview(k, bd = 100)=
{
  local(s,th,vk,wk,ck);

  th=sum(n=1,sqrtint(bd),2*x^(n^2),1+O(x^(bd+1)));
  vk=Vec(th^k-1);
  ck=Pi^(k/2)/gamma(k/2+1);
  wk=vector(bd);
  s=1;

  for(n=1,bd,

    s+=vk[n];
    wk[n]=s/ck/n^(k/2)-1);

  ploth(x=1,bd,wk[round(x)])
}
```

190   **8 : Remarks on selected exercises**

With this we get an idea of the size of $r_k(n)$ on average. See [50], Chapter 11 for more details on the actual size of $r_k(n)$. As is typical the sum $R_k(n) = \sum_{k=0}^{n} r_k(n)$ is much better behaved than $r_k(n)$ itself.

Ex. 2.4 This is a phenomenon first noticed by Chebyshev and is known as *Chebyshev's bias* after Rubinstein and Sarnak's paper [81].

Ex. 2.5 Here is a GP script that outputs a list of types and for which prime they first occur.

```
poltypesf(f,a = 2, b = 1000)=
{
  local(v,pv);

  w=[];pv=[];

  forprime(p=a,b,
    v=polfacttype(f,p);
      if(v==[] || memb(v,w),,
        w=concat(w,[v]);
        pv=concat(pv,p)));

  [pv,w]
}
```

To get all types for Trink's polynomial we do

```
? poltypesf(x^7-7*x+3,2,10^4)

[[2, 13, 17, 79, 1879], [[7], [4, 2, 1], [3, 3, 1],
              [2, 2, 1, 1, 1], [1, 1, 1, 1, 1, 1, 1]]]
```

Typically, the trivial type $[1,1,\ldots,1,1,1]$ requires a large p (compare with the Chebyshev bias of Ex. 2.4).

Ex. 2.9 This is one of a handful of isomorphisms that exist between classical groups (see [3]). Here is a list for the linear groups $\mathrm{PSL}_n(\mathbf{F}_q)$.

$\mathrm{PSL}_2(\mathbf{F}_2)$		S_3	6
$\mathrm{PSL}_2(\mathbf{F}_3)$		A_4	12
$\mathrm{PSL}_2(\mathbf{F}_4)$	$\mathrm{PSL}_2(\mathbf{F}_5)$	A_5	60
$\mathrm{PSL}_2(\mathbf{F}_7)$	$\mathrm{PSL}_3(\mathbf{F}_2)$		168
$\mathrm{PSL}_2(\mathbf{F}_9)$		A_6	360
$\mathrm{PSL}_4(\mathbf{F}_9)$		A_8	20160

where the last column is the order of the corresponding groups. Except for those in the first two rows the rest of the groups are simple. The group $PSL_3(F_4)$ has order 20160 but is *not* isomorphic to those in the last row.

Ex. 2.12 Here is a solution due to K. Belabas.

```
? a=x^7-7*x+3;
? b=x^8-4*x^7+7*x^6-7*x^5+7*x^4-7*x^3+7*x^2+5*x+1;
? compo(a,b) = local(v = polcompositum(a,b)); v[#v]
? K = compo(compo(b,b), b); \\ Galois closure of b,
                             \\ degree168.
? poldegree(compo(K, a))

168
```

Ex. 2.13 No. For example, in the group G of 3×3 upper triangular matrices with entries in F_3 and 1's along the diagonal every non-trivial element has order 3 but G is non-abelian.

```
? lift(Mod(1,3)*[1,a,b;0,1,c;0,0,1]^3)

[1 0 0]

[0 1 0]

[0 0 1]
```

Ex. 2.14 It is a consequence of (iii) of the following lemma (see [34], where the authors give an algorithm to compute the center of a Galois group).

Lemma 8.1 *Let $G \subseteq S(X)$ be a transitive subgroup acting on a set X. Let $Z(G)$ be the center of G. Then*
(i) The orbits of $z \in Z(G)$ acting on X have all the same size.
(ii) The center acts on X without fixed points.
(iii) The order $|Z(G)|$ of the center divides the number of fixed points in X of every $g \in G$.

Proof
(i) Let $u, v \in X$ be in two distinct orbits U and V of $z \in Z(G)$. Pick $g \in G$ such that $v = gu$ (the action is transitive). Then since $gzu = zgu = zv$ multiplication by g gives a bijection between U and V.
(ii) Follows immediately from (i).
(iii) If $gx = x$ then $zx = zgx = gzx$. Therefore $Z(G)$ acts on the fixed set of g. Now (iii) follows from (ii). □

Ex. 3.1 There is a couple of subtle implementation issues here.

(i) In bqf the `forstep` loop in b goes up to $b \leq \lfloor\sqrt{\lfloor|D|/3\rfloor}\rfloor$. On the other hand, the inequality we really get for a reduced form (a, b, c) is $|b| \leq \sqrt{|D|/3}$ (using that $\Im(z_Q) \geq \sqrt{3}/2$ for any point in the fundamental domain). Hence we need to verify that for any $x \geq 0$ we have $\lfloor\sqrt{\lfloor x\rfloor}\rfloor = \lfloor\sqrt{x}\rfloor$ (see Ex. 1.5).
The code

```
zv=divisors(z);
n=length(zv);

j=(n+1)\2;
a=zv[j];
c=zv[n-j+1];
```

does the following. We want to run over $ac = z$ where $z := (b^2 + D)/4$ and $b \leq a \leq c$. We compute all divisors of z with `zv=divisors(z)`. The divisors are now stored in the vector zv in increasing order. Therefore, the product of symmetrically located entries of zv (i.e., with indices j and $n-j+1$ for $1 \leq j \leq n$ with n is the length of zv, the number of divisors of z) is z. So once we set `a=zv[j]` we can just pick up c with `c=zv[n-j+1]` without having to perform the division $c = z/a$.
Also, we start at the middle of the vector zv so that a is as large as possible (recall $a \leq c$) and then decrease its value. In this way we know we can cut the loop off once $a < b$.

(ii) Questions regarding composition are typically tricky. You should at least check that bqfcomp does calculate the *Dirichlet composition* ([26] (3.7)) with the appropriate inputs. For a beautiful recent development on composition see [11].

Ex. 3.4 You should find only these discriminants:

$$-3, -4, -7, -8, -11, -12, -16, -19, -27, -28, -43, -67, -163.$$

It was a famous question of Gauss whether this list was complete. It was settled, independently, by Stark [90] and Baker [5] in the late 1960s. K. Heegner had in fact [47] published a proof in 1952, but was thought at the time to be flawed; Heegner's proof was not generally accepted until H. Stark published [91] his own account of it.

Ex. 3.5 This is part of what is called *genus* theory, see [26], Chapter I, §3 B.

Ex. 3.6 Here is a quick way to do the search. These discriminants are related to Euler's *idoneus numbers*, see [26], §3 C.

8.0 Remarks on selected exercises

```
idoneus(bd) =
{
  local(hv,h,j,S);

  S=[];
  for(d=1,bd,
    if(d%4==0 || d%4==3,
      h = qfbclassno(-d);
      if(h == 1,
         S=concat(S,[[-d,1]]),
         if(twopower(h),
      hv=bqf(-d);
      j=2;
      while(bqfpow(hv[j],2) == hv[1]
            && j < length(hv),j++);
      if(j == length(hv), S=concat(S,[[-d,j]])))))));
  S
}

twopower(n) =
{
  local(z,r);

  until(r,
    z=divrem(n,2);
    n=z[1];
    r=z[2]);
  (n==0)
}
```

The output is

```
? idoneus(10000)
[[-3, 1], [-4, 1], [-7, 1], [-8, 1], [-11, 1], [-12, 1],
 [-15, 2], [-16, 1], [-19, 1], [-20, 2], [-24, 2], [-27, 1],
 [-28, 1], [-32, 2], [-35, 2], [-36, 2], [-40, 2], [-43, 1],
 [-48, 2], [-51, 2], [-52, 2], [-60, 2], [-64, 2], [-67, 1],
 [-72, 2], [-75, 2], [-84, 4], [-88, 2], [-91, 2], [-96, 4],
 [-99, 2], [-100, 2], [-112, 2], [-115, 2], [-120, 4],
 [-123, 2], [-132, 4], [-147, 2], [-148, 2], [-160, 4],
 [-163, 1], [-168, 4], [-180, 4], [-187, 2], [-192, 4],
 [-195, 4], [-228, 4], [-232, 2], [-235, 2], [-240, 4],
 [-267, 2], [-280, 4], [-288, 4], [-312, 4], [-315, 4],
 [-340, 4], [-352, 4], [-372, 4], [-403, 2], [-408, 4],
 [-420, 8], [-427, 2], [-435, 4], [-448, 4], [-480, 8],
 [-483, 4], [-520, 4], [-532, 4], [-555, 4], [-595, 4],
 [-627, 4], [-660, 8], [-672, 8], [-708, 4], [-715, 4],
 [-760, 4], [-795, 4], [-840, 8], [-928, 4], [-960, 8],
 [-1012, 4], [-1092, 8], [-1120, 8], [-1155, 8], [-1248, 8],
 [-1320, 8], [-1380, 8], [-1428, 8], [-1435, 4], [-1540, 8],
 [-1632, 8], [-1848, 8], [-1995, 8], [-2080, 8], [-3003, 8],
 [-3040, 8], [-3315, 8], [-3360, 16], [-5280, 16],
 [-5460,16], [-7392, 16]]
```

It is conjectured that these are all there are but this is still an open problem. A different approach would be much faster: modify `bqf` so that it aborts as soon as we find a form which is not of order diving 2. However, it hardly seems worth the trouble (unless we really want to test the conjecture up to very large values of $|D|$).

Ex. 3.7 The point is that the fundamental domain of $SL_2(\mathbf{Z}[i])$ acting on hyperbolic 3-space is quite simple and yields an algorithm very close to that of Gauss. See [79] for some details. Here is a GP implementation closely following `bqfred`.

```
bhfred(H) =
{
  local(a,b,c,top,bot,n,aux);

  a=H[1];b=H[2];c=H[3];

  top=[-b,1]; bot=[shift(a,1),0];
  n=bqfn(a,real(b))+I*bqfn(a,imag(b));

  if(n,
     c+=a*norm(n)-shift(trace(n*conj(b)),-1);
     b-=shift(a*n,1);
     top+=n*bot);

  while(c < a,
     aux=c; c=a; a=aux;
     b=-conj(b);
     aux=top; top=-bot; bot=aux;
     n=bqfn(a,real(b))+I*bqfn(a,imag(b));

     if(n,
        c+=a*norm(n)-shift(trace(n*conj(b)),-1);
        b-=shift(a*n,1);
        top+=n*bot));

  if((imag(b) < 0) || (imag(b) == 0 && real(b) < 0),
     b=-b;
     top*=I;
     bot*=-I);

  if((a == c && real(b) < 0),
     b=-conj(b);
     aux=top; top=-bot; bot=aux;
```

```
H=[a,b,c];
bot[1]=(bot[1]-H[2]*bot[2])/2/H[1];

[H,bot]
}
```

Ex. 3.8   This is completely analogous to Cornachia's algorithm; we now start with a solution to the congruence

$$-1 \equiv u^2 + v^2 \bmod p$$

for example, computed with the following routine

```
sum2sqmod(p) =
{
  local(u,v);

  u=0; v=-1;

  while(kronecker(v,p) != 1,
    v-=shift(u,1)+1;
    u++);

  [u,lift(sqrt(Mod(v,p)))]
}
```

Then reduce the binary Hermitian form $H_p = (p, 2(u+iv), (u^2 + v^2 + 1)/p)$. Since there is only one reduced form of discriminant -4, namely $H = (1, 0, 1)$, in the process we find a solution to (3.9).

```
sum4sq(p) =
{
  local(H,v);
  v=sum2sqmod(p);
  H=bhfred([p,shift(v[1]+v[2]*I,1),(v[1]^2+v[2]^2+1)/p])[2];
  vecsort(abs([real(H[1]),imag(H[1]),real(H[2]),imag(H[2])]))
}
```

Though we have not checked the details it appears that sum4sq will actually output a solution to (3.9) with as many zero coordinates as is possible. In other words, if p is actually a sum of two or three squares (see §1.4.4) we will get such an expression for p. Here is a short list.

```
? forprime(p=2,50,v=sum4sq(p);print(p, "\t", v, "\t",
p-norml2(v)))
```

2	[0, 0, 1, 1]	0
3	[0, 1, 1, 1]	0
5	[0, 0, 1, 2]	0
7	[1, 1, 1, 2]	0
11	[0, 1, 1, 3]	0
13	[0, 0, 2, 3]	0
17	[0, 0, 1, 4]	0
19	[0, 1, 3, 3]	0
23	[1, 2, 3, 3]	0
29	[0, 0, 2, 5]	0
31	[1, 1, 2, 5]	0
37	[0, 0, 1, 6]	0
41	[0, 0, 4, 5]	0
43	[0, 3, 3, 5]	0
47	[2, 3, 3, 5]	0

Make sure to compare the running time of this algorithm with that of something like `sum4squares1` of §1.4.3.

Ex. 4.2 For (4.38) change variables and use the beta integral

$$\int_0^1 t^{\alpha-1}(1-t)^{\beta-1}dt = \frac{\Gamma(\alpha)\Gamma(\beta)}{\Gamma(\alpha+\beta)}, \quad \Re(\alpha), \Re(\beta) > 0,$$

where Γ is the usual gamma function (6.88), and

$$\frac{\Gamma(n+\frac{1}{2})}{\sqrt{\pi}\, n!} = \frac{1}{4^n}\binom{2n}{n} = (-1)^n \binom{-\frac{1}{2}}{n}.$$

Then (4.39) follows from (4.1).

Ex. 4.6 This example is from [2], §7; it is related to a point of order 8 on the elliptic curve $y^2 = x^3 - 5x^2 + 24x + 144$. There is a connection also to the 4-Somos sequences similar to those of Ex. 4.21 (see [93]).

Ex. 4.8 $\sum_{n=0}^{\infty} \binom{2n}{n} x^n = \frac{1}{\sqrt{1-4x}}, \quad \sum_{n=0}^{\infty} \frac{1}{n+1}\binom{2n}{n} x^n = \frac{1-\sqrt{1-4x}}{2x}.$

Ex. 4.9 This result is due to Lambert, see [75], p. 44.

Ex. 4.10 The numbers $C_{r,n}$ are generalizations of *Catalan numbers*, those with $r = 2$, and appear in many different combinatorial contexts, from which their integrality is obvious; for example in [89] Stanley gives 66 different descriptions of the standard Catalan numbers $C_{2,n}$ as the result of counting problems.

Ex. 4.12 Follows directly from Ex. 4.9 by differentiation. Hurwitz proved an analogous result for all rational r, see [75].

Ex. 4.13 This is a classical application of Lagrange's inversion formula. See also Polya's derivation from Lambert's series (4.54) in [75], p. 44.

Ex. 4.18 This sequence of numbers plays a key role in Apéry's proof of the irrationality of $\zeta(3)$, see [8].

Ex. 4.20 These two beautiful examples are connected to the work of Pisot [74] on his numbers. The first is due to D. Boyd [14], the second to D. Cantor [18], Table 5.6. Here is a somewhat lazy way to check Boyd's example.

```
cantorcheck(v,N)=
{
  local(a,aux);
  a0=3;a1=10;

  for(n=3,N-1,
    aux=floor(a1^2/a0+1/2);
    a0=a1;
    a1=aux;
    if(v[n+1]!=a1,print(n)));
}
```

Here we input the vector of coefficients of the rational function and $N = 11057$. WARNING: The numbers get really big and hence we need to allocate enough memory (with `allocatemem`) for PARI to be able to do the calculations; it also takes a while to perform the full test.

```
? R=(8+7*x-7*x^2-7*x^3)/(1-6*x-7*x^2+5*x^3+6*x^4);
? v=Vec(R+O(x^11057));
? boydcheck(v,11057)

11057
```

Similarly for Cantor's example

```
cantorcheck(v,N)=
{
  local(a,aux);
  a0=3;a1=10;

  for(n=3,N-1,
    aux=floor(a1^2/a0+1/2);
    a0=a1;
    a1=aux;
    if(v[n+1]!=a1,print(n)));
}
```

where now we do

```
? v=Vec(1/(1-x*(x+3))+O(x^1000));
? cantorcheck(v,1000)
```

which should have no output. As usual, of course, this does not *prove* the identity!

Ex. 4.21 This is an example of a sequence that satisfies a bilinear recursion, rather than linear, which has received considerable attention recently; they are generically known as Somos sequences (after M. Somos who first considered them).

The fact that the 6-Somos sequence is always integral has been proved by Hickerson and Stanley (independently) using computer algebra. More recently Fomin and Zelevinsky [35] using their notion of a cluster algebras prove this and a number of other similarly amazing facts. They show that certain recursions of polynomials in many variables unexpectedly produce only Laurent polynomials; in particular, setting all variables to 1 gives a sequence of integers. This is known as the *Laurent phenomenon*.

For a brilliant analysis of a similar 5-Somos sequence, Sloane A006721, see [100]. The sequence is connected to addition on an elliptic curve. Check the Somos sequence website for more information:

http://www.math.wisc.edu/propp/somos.html

Ex. 4.22 This is an instance of Guy's *law of small numbers* [44], example 24. The first case where a_n not integral occurs for $n = 43$ where $a_n \approx 5.4093 \times 10^{178485291567}$ so direct verification is not feasible, see [100]. No later term is integral after that.

Ex. 4.23 Check the original papers [96] and [31] for information on these sequences.

Numerically it is better to consider the quantities $r_n := x_{n+1}/x_n$ to keep the terms in the recursion of a reasonable size. These satisfy the recurrence

$$r_{n+1} = 1 \pm \frac{\beta}{r_n}$$

and

$$\lambda_n = \frac{1}{n} \sum_{k=0}^{n-1} \log |r_k|$$

which we can compute with, for example,

```
randfib(m,b) =
{
  local(f0,f,s,v);
  v=vector(m+1);
  s=0;v[1]=1;f0=1;
  for(n=1,m,
    f=1+b*(-1)^random(2)/f0;
    s=s+log(abs(f));
    v[n+1]=s/(n+1);f0=f);
  v
}
```

The output is a vector with all the values $\lambda_0, \lambda_1, \ldots, \lambda_m$. We can then plot them to see their behavior with, say,

```
plv(v)= ploth(x=1,length(v),v[round(x)])
```

Remarkably, λ_n does indeed have a limiting value (with probability one) $\lambda = \lambda(\beta)$ as $n \to \infty$ as the numerical experimentation suggests.

Ex. 4.25 (M. Newman) The sequence x_n is an enumeration of all positive rational numbers. This is related to the Stern–Brocot tree, Eisenstein–Stern diatomic sequence, Sloane A002487, and the Calk–Wilf tree (see [19]).

Ex. 4.26 If $y = M(x)$ then $x = 2\lceil \frac{1}{y} \rceil - 1 - \frac{1}{y}$.

Ex. 4.27 The sequence reproduces (see [12]) the notorious, and still open, $3x + 1$ problem (see [55] for a survey): for any starting value $x_1 \in \mathbb{Z}$ the sequence should eventually reach 1. Here is a GP script for a quick check

```
syracuse(a)=
{
  local(v);
  v=[a];

      while(a>1,a=(1+a)*a-floor(a/2)*(2*a+1);v=concat(v,a));
  v
}
```

and a sample output

```
? syracuse(100)

[100, 50, 25, 38, 19, 29, 44, 22, 11, 17, 26, 13, 20, 10,
5, 8, 4, 2, 1]
```

For a fast way to check the conjecture for large inputs see [95], Chapter 7.

Ex. 5.6   There are several possibilities. One can use the fact that
$$d(n) = \left[\frac{n!}{e}\right],$$
(here $[\cdot]$ is the nearest integer function), the linear recursion
$$d(n) = (n-1)(d(n-1) + d(n-2)),$$
or
$$d(n) = nd(n-1) + (-1)^n.$$
The first few values are $0, 1, 2, 9, 44, 265, 1854, \ldots$ (Sloane A000166).

Ex. 5.7   Note that by (5.8) $d_n/\sqrt{n!}$ is a natural normalization with $d_n/\sqrt{n!} < 1$. See [51], Theorem A, p. 110.

Ex. 5.8   This is tricky. The order r of an element in S_n only depends on its conjugacy class (i.e., its cycle decomposition) which is indexed by a partition $(\lambda_1, \lambda_2, \ldots)$ of n and, in fact, $r = \text{lcm}(\lambda_1, \lambda_2, \ldots)$. We can easily write a routine using the ideas of this chapter to run over partitions of $k \leq n$ while computing $r = \text{lcm}(\lambda_1, \lambda_2, \ldots)$ to find the maximum. However, this will *very* inefficient. One should instead use that
$$g(n) = \max \prod_i p_i^{r_i}, \quad \sum_i p_i^{r_i} = n \qquad (8.1)$$
with p_i distinct primes but we have not attempted it.

As a further challenge we leave to the reader the analysis of why the following GP function comes close but fails to compute $g(n)$ in general.

```
landaux(n) =
{
  local(v);

  if(n<0,0,
    n++;v=vector(n,i,1);
    for(j=2,n,
      for(i=1,j-1,
        v[j]=max(v[j],lcm(i,v[j-i]))));
    v)
}
```

Landau [57] proved that $\log g(b) \sim \sqrt{n \log n}$ as $n \to \infty$, see also [69].

Ex. 5.11   Hurwitz used this calculation to count certain ramified covers of the Riemann sphere [48].

Ex. 5.17   Here is one possibility.

```
nondecseq(n,a,b)=
{
  local(k,j,sj,S);
  S=[];
  k=1;j=a;
  sj=vector(n+1);

  while(k,

       if(k > n,
          until(j <= b, j=sj[k]+1; sj[k]=0; k--);
          next);

       k++;   sj[k]=j;

       S=concat(S, [vector(k-1,1,sj[l+1])]));

  S
}
```

Ex. 6.1   See [101].

Ex. 6.4   The only way we know how to prove this is by checking that $v_p(a_n) \geq 0$ for all n and all primes p. No combinatorial interpretation of this numbers is known. This fact was used by Cheyshev in his work on the number of prime numbers less than a given bound. The generating series $\sum_{n\geq 0} a_n x^n$ is an algebraic function of degree 483, 840, see [80].

Ex. 6.8   This result is due to Jacobsthal and generalizes (5.26). It is a special case of (5.19).

Ex. 6.18   This concept arose in the work of Kummer on Fermat's last theorem; more precisely, Kummer proved that $x^p + y^p = z^p$ has no solutions in positive integers if p is regular (see [49] 233–234). For a large scale computation of irregularity see [16] and the references therein.

Ex. 7.3   These sequences are the main focus of Lehmer's paper [58], where he raises his famous question, see §7.1. Lehmer was interested in using them to find large prime numbers. The statements (i)–(iii) are proved there.

To find a recursion for $\Delta_n(P)$ (not necessarily the one of smallest degree) we can follow Lehmer ([58] §8 (though he seems to have the recursion in his theorem backwards)). We compute the least common multiple of $f_0, f_2, \ldots, f_8$, where f_k is the polynomial whose roots are the products of k-distinct roots of f. To compute this we use $G_r f$ as explained in §7.2.3.

Here is the calculation (plus a check) for the given polynomial. We use that $\Delta_n(P) = (-1)^{n \deg(P)} \operatorname{Res}(x^n - 1, P)$. Compare with [58] §14.

```
? f=x^8-x^5-x^4-x^3+1;
? wv=polwedge(f,8);
? h=1;for(k=1,length(wv),h=lcm(h,wv[k]));
? poldegree(h)

126
? delv=vector(500,n,polresultant(x^n-1,f));
? reccheck(k,v,pol)=local(m);m=poldegree(pol);
            sum(n=0,m,v[k+n]*polcoeff(pol,m-n))
? vector(20,k,reccheck(k+300,delv,11))

[0, 0, 0, 0, 0, 0, 0, 0, 0, 0, 0, 0, 0, 0, 0, 0, 0, 0, 0, 0]
? delv[55]

-529
? delv[500]

-35265789053998927963614182824199526865847598836389083
```

Ex. 7.4 About 88% of the polynomials in Boyd's list with Mahler measure less than 1.3 and degree at most 32 are of this form, see [68], §3.

Ex. 7.7 See [63] for the growth of the coefficients of the n-th cyclotomic polynomial Φ_n. A simple check in GP could be this

```
? n=1;until(polcyclomax(n)>1,n++);print(n)

105
```

where we used the function

```
polcyclomax(n)=vecmax(abs(Vec(polcyclo(n))))
```

to compute the largest coefficient (in absolute value) of Φ_n.

A longer search reveals the following list of n's where the maximum goes up (you will probably need to increase your stack with `allocatmem()` and wait a while to reproduce this).

```
? m=1;for(n=1,3*10^4,mn=polcyclomax(n); if(mn > m,
  print(n,"\t",mn); m=mn))
```

105	2
385	3
1365	4
1785	5
2805	6
3135	7
6545	9
10465	14
11305	23
17255	25
20615	27
26565	59

Ex. 7.8 The answer to this exercise gives essentially *all* interlacing cyclotomic polynomials by a theorem of Beukers and Heckman [9]. Essentially here means up to the obvious changes $x \mapsto x^k$ for some $k > 1$ and $x \mapsto -x$ and apart from the infinite family of examples $p = (x^n - 1)/(x - 1), q = (x^r - 1)(x^s - 1)/(x - 1)$ where $r + s = n$. The example $\Phi_{30}, \Phi_1\Phi_2\Phi_3\Phi_5$ is related to the Chebyshev numbers a_n of Ex. 6.4 (see also [80]).

Ex. 7.9 Here is a way to do it.

```
ser2prod(f,m) =
{
   if(m==0,m=length(f)+1);
   f=f+O(x^m);
   if(subst(truncate(f),x,0)!=1,
   error("Series must be of the form 1 + ..."),
   f=-log(f);
   sum(n=1,length(f),
    -sumdiv(n,d,polcoeff(f,d)*moebius(n/d)*d)/n*x^n))
}
```

The answer is given in the form $\sum_{n\geq 1} c_n x^n$. Here is an example

```
? f=polcyclo(9)^2*polcyclo(3)*polcyclo(12)*polcyclo(4);
? g=ser2prod(f)

-x^12 - 2*x^9 + x^6 + x^3 + x
? prod(k=1,poldegree(g),(1-x^k)^(-polcoeff(g,k)))-f

0
```

References

[1] G. E. Andrews, *Euler's 'exemplum memorabile inductionis fallacis' and q-trinomial coefficients*, J. Amer. Math. Soc. **3** (1990) 653–669.

[2] W. Adams, M. J. Razar, *Multiples of points on elliptic curves and continued fractions*, Proc. London Math. Soc. **41** (1980) 481–498.

[3] E. Artin, *The orders of the classical simple groups*, Comm. Pure Appl. Math. **8** (1955), 455–472; Collected papers, Addison-Wesley, Reading, MA 1965, p. 398.

[4] B. Baillaud and H. Bourget, *Correspondance d'Hermite et de Stieltjes*, I,II, Gauthier-Villars, Paris 1905.

[5] A. Baker, *Linear Forms in the Logarithms of Algebraic Numbers. I.*, Mathematika **13**, (1966) 204–216.

[6] M. Beeler, R. W. Gosper, R. Schroeppel, HAKMEM
http://www.inwap.com/pdp10/hbaker/hakmem/hakmem.html

[7] K. Belabas
http://www.math.u-bordeaux.fr/belabas/teach/2005/MAT913/book.pdf

[8] F. Beukers, *A note on the irrationality of $\zeta(2)$ and $\zeta(3)$*, Bull. London Math. Soc. **11**, (1979) 268–272.

[9] F. Beukers and G. Heckman, *Monodromy for the hypergeometric function $nFn-1$*, Invent. Math. **95** (1989) 325–354.

[10] F. Beukers and C. J. Smyth, *Cyclotomic points on curves*, Number Theory for the Millenium I, A. K. Peters 2002 (Proceedings of the Millennial Conference on number theory, Urbana May 21–26 2000), 67–85.

[11] M. Bhargava, *Higher composition laws I: A new view on Gauss composition, and quadratic generalizations*, Ann. of Math. **159**, no. 1 (2004).

[12] E. Bombieri and A. van der Poorten, Continued fraction of algebraic numbers, Computational algebra and number theory (Sydney, 1992), pp. 137–152.

[13] D. Boyd, *Reciprocal polynomials having small Mahler measure*, Math. Comp. **35** (1980) 1361–1377.

[14] D. Boyd, *Pisot sequences which satisfy no linear recurrence*, Acta Arith., **32** (1977) 89–98.

[15] D. Boyd, *A p-adic study of the partial sums of the harmonic series*, Experim. Math. 3 (1994) 287–302.

[16] J. Buhler, R. Crandall, R. W. Sompolski, *Irregular primes to one million*, Math. Comp. **59** (1992) 717–722.

[17] P. Candelas, X. de la Ossa and F. Rodriguez Villegas, *Calabi–Yau manifolds over finite fields. II* Calabi–Yau varieties and mirror symmetry (Toronto, ON, 2001), 121–157, Fields Inst. Commun., 38, Amer. Math. Soc., Providence, RI, 2003.

[18] D. G. Cantor, *On families of Pisot E-sequences*, Ann. Sci. Ecole Norm. Sup., **9** (1976) 283–308.

[19] N. J. Calkin and H. S. Wilf *Recounting the rationals*, Amer. Math. Monthly, **107** (2000) 360–363.

[20] J. W. S. Cassels *Local Fields* LMS, Student Texts 3, Cambridge University Press 1986.

[21] J. W. S. Cassels and A. Frölich (eds.), *Algebraic Number Theory*, Academic Press 1967.

[22] H. Cohen, *A Course in Computational Algebraic Number Theory*, **138** GTM Springer-Verlag, 1993.

[23] H. Cohen, *Algebraic Number Theory*, Springer Verlag (to appear).

[24] G. Collins and M. Encarnación, *Efficient rational number reconstruction* J. Symbolic Comput. 20 (1995) 287–297.

[25] L. Comtet, *Calcul pratique des coecients de Taylor d'une d'une fonction algbrique*, Enseignement Mathématique. **10** (1964) 267–270.

[26] D. Cox, *Primes of the form $x^2 + ny^2$*, John Wiley & Sons, NewYork, 1989.

[27] D. Cox, *The arithmetic-geometric mean of Gauss*, Enseign. Math. 30 (1984), 3–4, 275–330.

[28] J. Diamond, *The p-adic log gamma function and p-adic Euler constants*, Trans. Amer. math. Soc., **233** (1977) 321–337.

[29] B. Dwork, *p-adic cycles*, Pub. Math. I.H.E.S. 37 (1969) 27–116.

[30] O. Egecioglu and C. K. Koc, *A fast algorithm for rational interpolation via orthogonal polynomials*, Math. Comp., **53** (1989) 249–264.

[31] M. Embree and L. N. Trefethen *Growth and decay of random Fibonacci sequences*, Proc. Roy. Soc. London A **455** (1999) 2471–2485.

[32] L. Euler, *Exemplum Memorabile Inductionis Fallacis*, Opera Omnia, Series Prima, **15**, Teubner, Leipzig und Berlin, (1911) 50–69.

[33] R. J. Fateman, *Algorithms for manipulating formal power series* JACM **25** (1978) 581–595.

[34] P. Fernandez-Ferreiros and M. A. Gomez-Molleda, *Deciding the nilpotency of the Galois group by computing elements in the centre*, Math. Comp. **73** (2004) 2043–2060.

[35] S. Fomin and A. Zelevinski, *The Laurent phenomenon*, Adv. in Appl. Math. **28** (2002) 119–144.

[36] J. S. Frame, F. de Robinson and R. M. Thrall *The hook graphs of the symmetric group* Canad. J. Math. **6** (1954) 316–324.

[37] W. Fulton and J. Harris. *Representation Theory*. Springer, 1991.

[38] *GAP. Groups, Algorithms, Programming, a System for Computational Discrete Algebra* http://www-groups.dcs.st-andrews.ac.uk/gap/gap.html

[39] F. Gassman, *Bemerkungen zur vorstehenden Arbeit von Hurwitz*, Math. Z., **25** (1926) 124–143.

[40] H. W. Gould *Explicit formulas for Bernoulli numbers*, Amer. Math. Monthly **79** (1972) 44–51.

[41] C. Gordon, D. L. Webb, and S. Wolpert, *One cannot hear the shape of a drum* Bull. AMS, **27** (1992) 134–138.

[42] R. L. Graham, D. E. Knuth, and O. Patashnik, *Concrete Mathematics*, Reading, MA, Addison-Wesley, 1994.

[43] B. Gross and N. Koblitz, *Gauss sums and the p-adic Γ-function*, Ann. of Math. **109** (1979) 569–581.

[44] K. Guy, *The strong law of small numbers*, Amer. Math. Monthly **95** (1988) 697–712.

[45] F. Hajir and F. Rodriguez Villegas, *Explicit elliptic units*, Duke Math. J. **90**, (1997) 495–521.

[46] G. H. Hardy and E. M. Wright, (1960) *An Introduction to the Theory of Numbers*, 4th edn., Oxford University Press, Oxford.

[47] K. Heegner, K. *Diophantische Analysis und Modulfunktionen*, Math. Z. **56** (1952) 227–253.

[48] A. Hurwitz, Über die Anzahl der Riemann'schen Flächen mit gegebenen Verzweigungspunkten, Math. Ann. **55** (1902) 53–66.

[49] K. Ireland, M. Rosen, *A Classical Introduction to Modern Number Theory*, 2 edn, Springer-Verlag, New York (1990).

[50] H. Iwaniec, *Topics in Classical Automorphic Forms*, Grad. Studies in Mathematics, **17** AMS, Providence, RI, 1997.

[51] S. V. Kerov *Asymptotic Representation Theory of the Symmetric Group and its Applications in Analysis*, Translations of Mathematical Monographs **210**, AMS, Providence RI 2003.

[52] D. Knuth, *The Art of Computer Programming*, Reading, Massachusetts, Addison-Wesley (1997).

[53] L. Kronecker *Über die Bernoulli Zahlen,* in Mathematische Werke vol. II, (1883) 403–407.

[54] H. T. Kung and J. F. Traub *All algebraic functions can be computed fast,* JACM **25** (1978) 245–260.

[55] J. Lagarias *The 3x+1 problem and its generalizations,* Amer. Math. Monthly, **92** (1985) 3–23.

[56] J. Lagarias and A. Odlyzko, *Effective versions of the Chebotarev density theorem,* in Algebraic number fields: L-functions and Galois properties (Proc. Sympos., Univ. Durham, Durham, 1975) 409–464. Academic Press, London, 1977.

[57] E. Landau, *Handbuch der Lehre von der Verteilung der Primzahlen,* B. G. Teubner, Leipzig und Berlin, (1909).

[58] D. H. Lehmer, *Factorization of certain cyclotomic functions,* Ann. of Math. (2) **34** (1933) 461–479.

[59] D. H. Lehmer, *A new approach to Bernoulli polynomials,* Amer. Math. Monthly **95** (1988) 905–911.

[60] A. K. Lenstra, H. W. Lenstra L. Lovasz, *Factoring polynomials with rational coefficients,* Math. Ann. **261** (1982) 515–534.

[61] K. Mahler, *On some inequalities for polynomials in several variables,* J. London Math. Soc. **37** (1962) 341–344.

[62] K. Mahler, *Lectures on Diophantine approximations,* 1, University of Notre Dame (1961).

[63] H. Maier, *The coefficients of cyclotomic polynomials,* Analytic number theory (Allerton Park, IL, 1989), 349–366, Progr. Math., 85, Birkhäuser Boston, Boston, MA, 1990.

[64] W. H. Mills, *Continued fractions and linear recurrences,* Math. Comp., **29** (1975) 173–180.

[65] E. Mortenson, *A supercongruence conjecture of Rodriguez-Villegas for a certain truncated hypergeometric function,* JNT **99** (2003) 139–147.

[66] Mossinghoff, *Computational Aspects of Problems on Mahler's Measure,* a talk for a short graduate course at the PIMS Workshop on Mahler's Measure of Polynomials, Simon Fraser University, June 2003.

[67] M. J. Mossinghoff, *Lehmer's Problem,* http://www.cecm.sfu.ca/mjm/Lehmer/

[68] M. J. Mossinghoff, C. G. Pinner and J. D. Vaaler, *Perturbing polynomials with all their roots on the unit circle,* Math. Comp. **67** (1998) 1707–1726.

[69] J. L. Nicolas, *Ordre maximal d'un élément du groupe S_n des permutations et 'highly composite numbers',* Bull. de la Soc. Math. de France, **97** (1969) 129–191.

[70] A. M. Odlyzko, *Asymptotic enumeration methods*, in Handbook of Combinatorics, vol. 2, R. L. Graham, M. Groetschel, and L. Lovasz, eds., Elsevier, 1995 1063–1229.

[71] R. Perlis, *On the equation $\zeta_K(s) = \zeta_{K'}(s)$*, J. Number Theory, **9** (1977) 342–360.

[72] R. Perlis, *Thinking Disconnectedly about Klein's 4-group*, http://www.math.lsu.edu/perlis/2Stark65.pdf

[73] M. Petkovsek, H. Wilf, D. Zeilberger, (1996) A = B, A. K. Peters, Wellesley, MA.

[74] Ch. Pisot, *La répartition modulo 1 et les nombres algébriques*, Ann. Scuola Norm. Sup. Pisa Cl. Sci., **7** (1938) 205–248.

[75] G. Polya, *Sur les séries entières dont la somme est une function algébrique*, Enseig. Math. **1–2** (1921–1922) 38–47.

[76] B. Poonen and M. Rubinstein, *The number of intersection points made by the diagonals of a regular polygon*, SIAM J. Discrete Math. **11** (1998) 135–156.

[77] H. Rademacher, *On the Expansion of the Partition Function in a Series*, Ann. Math. **44** (1943) 416–422.

[78] A. Robert, *The Gross–Koblitz formula revisited*, Rend. Sem. Mat. Univ. Padova, **105** (2001) 157–170.

[79] F. Rodriguez Villegas, *Explicit models of genus 2 curves with split CM* Algorithmic number theory (Leiden, 2000), 505–513, Lecture Notes in Comput. Sci., **1838**, Springer, Berlin, 2000.

[80] F. Rodriguez Villegas, *Integral ratios of factorials and algebraic hypergeometric functions*, Oberwolfach report (2005).

[81] M. Rubinstein and P. Sarnak, *Chebyshev bias*, Experim. Math. **3** (1994) 173–197.

[82] W. H. Schikhof, *Ultrametric Calculus: An Introduction to p-adic Analysis*, Cambridge University Press, Cambridge (1984).

[83] J.-P. Serre, *A Course in Arithmetic*, GTM 7, Springer-Verlag, New York, 1973.

[84] J.-P. Serre, *Modular Forms and Galois Representations*, Collected papers III, Springer Verlag, Berlin Heidelberg 1986.

[85] J.-P. Serre, *Topics in Galois Theory*, Research Notes in Math. **1**, Jones and Bartlett Publisherse, Boston, MA, 1992.

[86] N. J. Sloane, *The On-Line Encyclopedia of Integer Sequences* http://www.research.att.com/njas/sequences/

[87] C. J. Smyth, *On measures of polynomials in several variables* Bull. Austral. Math. Soc. **23** (1981) 49–63.

[88] R. Stanley, *Enumerative Combinatorics*, Vol. 1, Wadsworth & Brooke, Monterey, CA.

[89] R. Stanley, *Enumerative Combinatorics*, Vol. 2, Cambridge Univ. Press, Cambridge.

[90] H. M. Stark, *A complete determination of the complex quadratic fields of class number one*, Michigan Math. J. **14** (1967) 1–27.

[91] H. M. Stark, *On the "gap" in a theorem of Heegner*, J. Number Theory **1** (1969) 16–27.

[92] T. Sunada, *Riemannian coverings and isospectral manifolds*, Ann. of Math., **121** (1985) 169–186.

[93] A. van der Poorten, *Elliptic curves and continued fractions*, J. Int. Seq. **8** (2005). Article 05, 2.5

[94] L. van Hamme, *Some congruences involving the p-adic gamma function and some arithmetical consequences*, in *p-adic functional analysis (Ioannina, 2000)*, (2001) 133–138, Lecture Notes in Pure and Appl. Math., **222**, Dekker, New York.

[95] I. Vardi, *Computational Recreations in Mathematica*, Addison-Wesley, 1991.

[96] D. Viswanath, *Random Fibonacci sequences and the number 1.13198824...*, Math. Comp. **69** (2000) 1131–1155.

[97] P. Wang, M. J. T. Guy, J. H. Davenport, *p-adic reconstruction of rational numbers*, SIGSAM Bulletin, **16**, No 2 (1982) 2–3.

[98] P. G. Walsh, *A polynomial-time complexity bound for the computation of the singular part of a Puiseux expansion of an algebraic function* Math. Comp. **69** (2000) 1167–1182.

[99] D. Zagier, *A one-sentence proof that every prime $p \equiv 1 \pmod{4}$ is a sum of two squares*, Amer. Math. Monthly **97** (1990) 144.

[100] D. Zagier, *Problems posed at the St Andrews Colloquium, 1996*, http://www-groups.dcs.st-and.ac.uk/john/Zagier/Problems.html

[101] D. Zagier, J. Shallit, N. Strauss, *Problems and Solutions: 6625*, Amer. Math. Monthly **99** (1992) 66–69.

Index

Artin
   map, 50, 86, 90
   reciprocity law, 40, 42, 65, 71, 77, 78, 80
   representation, 81

Baker, A., 192
Belabas, K., 32, 38, 102, 175, 191
Bernoulli
   number, 11, 20, 21, 108, 165, 167, 176
   polynomial, 19, 20, 23, 24
Bernoulli, J., 23
bernoulli, 20
Beukers, F., 203
Binary Hermitian form, 91
Boyd, D., 162, 179–181, 197
boydcheck, 197
bqf, 75
bqfcomp, 76
bqfistrivial, 79
bqfpow, 77
bqfprimef, 79
bqfred, 72
bqftheta, 80

Calf–Wilf tree, 199
cantorcheck, 197
Catalan number, 119, 196
centerliftp, 174
Chebyshev
   bias, 190
checkbqf, 75
checkfermat1, 74
clogp, 160
coefficients of algebraic functions, 176
Comtet, L., 99
contfracper, 106
contfracrq, 116
Cornachia algorithm, 32, 73, 78
count, 4

de Smit, B., 189
Dedekind η function, 17–19, 129

Delannoy number, 120
diam, 163
Diamond, J., 163, 176
diamtest, 164
Dirichlet
   character, 1, 4–6, 44
   class number formula, 163
   composition, 192
dist, 7
Dwork
   character, 172, 173
   exponential, 170
Dwork, B., 157, 168, 169

Eisenstein, *see* coefficients of algebraic functions
Eisenstein–Stern diatomic sequence, 199
ellunit, 18
enumerating the rationals, 199
Exemplum Memorabile Inductionis Fallacis, 121
Euler, 10, 11, 92, 129
   ϕ function, 123, 143
   constant, 163
   criterion, 169
   factor, 43, 44, 65
   idoneus number, 192
   infinite product, 129
   partition formula, 148
eulerphilist, 144

Fermat
   last theorem, 201
   sum of two squares, 32
fermat1, 73
Fibonacci
   random sequence, 122
   sequence, 121
floorsqrt, 15
Fomin, S., 198
Fourier
   fast transform, 36

Frobenius, 138
  number of homomorphisms, 149
  automorphism, 50, 88
  conjugacy class, 5, 55
Frobenius–Schur formula, 149
Frobenius–Hurwitz formula, 148, 149

gammap, 156
gammapv, 157
Gassman, F., 65, 69
Gauss
  ψ formula, 163
  class number one, 192
  class number, sum of three squares, 38
  composition, 76
  Disquitiones Arithmeticae, 71
  quadratic symbol lemma, 159
  reduction, 32, 75
  sum, 172–174
Geldon, T., 134
Glaisher's congruence, 176
Göbel sequence, 122
gosper, 13
Graeffe
  general case, 182
  root-squaring method, 180
Gross–Koblitz formula, 172–174, 176
gsumc, 174

harmp, 162
Hecke, *see* dihedral representation
Heckman, G., 203
Heegner, K., 192
Hensel's lemma, 99, 100, 103, 117, 120, 152, 153
hensel, 100
hensel1, 117
Hickerson, D., 198
Hurwitz
  algebraic function, 196
  quaternions, 33

idoneus, 192
interlacing, 186

Jacobi symbol, 159

Klein's order 168 group, 60, 142
Krasner, M., 168
Kronecker
  symbol, 2, 4, 6, 48
  theorem, 143, 177
Kronecker, L., 20
Kronecker–Weber theorem, 43
Kummer, E., 201

Lagrange
  interpolation, 114
  inversion formula, 119, 196
  sum of four squares theorem, 27, 29, 31
Lambert series, 196, 197
Landau function, 148, 200
landaux, 200
Laplace method, 108
Laurent
  expansion, 106
  phenomenon, 198
  polynomial, 92, 98, 198
  series, 103–105
Legendre
  polynomials, 107
  symbol, 169
Lehmer, 201
  polynomial, 178, 179
  question, 178
  recursion, 201
  sequence, 201
Lehmer question website, 178
Lenstra, H., 15
lineq, 125
LLL, 15
logeps, 163
Lovasz, L., 15

Mahler
  expansion, 151, 155, 156
  measure, 177, 178
mahlerlist, 178
money changing problem, 127
Mortenson, E., 9
Mossinghoff, M., 179
Motzkin sequence, 120
Murnaghan–Nakayama rule, 133, 136

Newman, M., 199
Newton
    formula, 185
    method, 99, 101, 103, 152–154
newtsqrt, 102
nondecseq, 201

Padé approximation, 102, 107
part, 127
partdim, 132
partdual, 131
partition, 41, 50, 126, 127
partnum, 128
partnum1, 129
partnum2, 130
partnum3, 130
partnum4, 130
parttglhom, 140
pentagonal number, 129
Perlis, R., 69
permno, 34
permno1, 38
Pisot number, 197
polcycloexp, 187
polcyclofactor, 184
polcyclolist, 145
polfacttype, 44
polgal8, 60
polgalperm, 55
poliscyclo, 183
polisgalois, 60
pollisttype, 45
pollisttype1, 46
polsum, 23
poltypedensity1, 58
poltypes, 45
poltypesf, 190
polwedge, 185
psip, 161
$PSL_2(F_7)$, see Klein's order 168 group

Rademacher's series, 130
Rademacher's series, 128
Ramanujan taxicab number, 39
randfib, 198
recognize, 11
recognize1, 14
recognize2, 13
recognizemod, 15
recsolve, 97
Riemann
    sphere, 200
    surface, 149

zeta function, 10, 44
Rubinstein, R., 190
Runge, C., 168

Sarnak, P., 190
Schur, see Frobenius–Schur formula
seqisratnl, 115
seqlinrec, 96
ser2prod, 203
seralgdep, 99
serdiffeq, 94
serlindep, 94, 115
sgn, 7
sgn1, 9
sgn2, 10
signsno, 34
Somos, M., 198
Somos sequence, 121, 196, 198
somos6, 121
ssasymptview, 189
Stanley, R., 196, 198
Stark, H., 192
Stephan, R., 130
Stern, see Eisenstein–Stern diatomic
    sequence
Stern–Brocot tree, 199
Stickelberger theorem, 173
Stirling's series, 108
Stirling formula, 108
Sturm method, 182
sum2squares1, 29
sum2squares2, 31
sum2squares3, 32
sum2squares4, 32
sum3squares1, 29
sum4squares1, 30
sum4squares2, 31
sumasympt, 112
sumsq, 33
sumsquares, 27
Sunada, T., 65, 69
syracuse, 199

Tate, J., 75
Taylor
    coefficient, 99
    expansion, 120, 121
    theorem, 25, 26, 100, 101
Tchebotarev density theorem, 49, 55, 56,
    60, 63, 65, 69
test, 79
testmod, 6
trinapprox, 112

`trinasympt`, 110
Trink polynomial, 60, 64, 65, 69, 190
`trinomial`, 92

Volkenborn integral, 164
Voloch example, 84

Wolstenholme congruence, 166, 176

Zagier, D., 20
Zelevinsky, A., 198